The Riot and the Dance
Lab Manual

THE RIOT AND THE DANCE
LAB MANUAL

DR. GORDON WILSON

Gordon Wilson, *The Riot and the Dance Lab Manual*

Published by Canon Press
P.O. Box 8729, Moscow, Idaho 83843
800.488.2034 | www.canonpress.com

Cover design by James Engerbretsen. Cover illustrations by Forrest Dickison.
Interior design by Laura Storm Design. Interior Layout by Valerie Anne Bost
Printed in the United States of America.

18 19 20 21 22 10 9 8 7 6 5 4 3 2

CONTENTS

PUBLISHER'S NOTE

This is the Lab Manual for *The Riot and the Dance,* a new biology text from Dr. Gordon Wilson that focuses on teaching students the integrated fundamentals of biology in an approachable and yet detailed way. The Lab Manual is an important tool to show students how the concepts they are learning relate to real life. As such, you'll use a lot of "real life" materials. Each lab begins with a clearly labeled "Materials" list, many of which should be available in your kitchen or grocery store. For materials that must be specially ordered, we recommend one of these online supply sources: Carolina Biological Supply Company, Bio Corporation, and Ward's Science. Make sure you give yourself enough time to receive these materials in the mail! Some labs also require some slight preliminary preparation, so be sure to check the "Preparation" section of each lab ahead of time as well.

Several labs call for students to watch videos online on YouTube or other sites—the twenty-first century version of classroom film strips! To save having to type in all of those URLs, we've prepared a list of all the video links you will need for the course: http://logospressonline.com/content/RiotLinks.pdf. You'll probably want to bookmark this in your browser for easy access. You can contact us online (www.logospressonline.com) or by phone (208.892.8074) if you have any questions.

The goal of this Lab Manual is to stir up curiosity about all of life from cells to sharks to ecology—along with a greater desire to praise the Creator of it all. Enjoy!

SUGGESTED YEAR-LONG SCHEDULE

Below is a suggested schedule for working through The Riot and the Dance Textbook, Lab Manual, and Teacher's Guide, meeting five days a week over two semesters. If you meet fewer times per week, condense the schedule as needed. Each week will have at least two days of teaching and reading through that week's material, along with review, weekly quizzes & exams, and 25 labs.

WEEK	DAY	LECTURE/LABORATORY
1	1	Introduction
	2	Begin Ch. 1: A Smidge of Chemistry
	3	Finish Ch. 1
	4	Review Questions for Quiz
	5	Ch. 1 Quiz (Ch. 1 review)
2	1	Begin Ch. 2: Biomolecules
	2	Finish Ch. 2 & Review Questions
	3	***Lab 1: The Microscope****
	4	Ch. 2 Quiz & review for Exam
	5	Unit 1 Exam
3	1	Begin Ch. 3: A Short History of Microscopy
	2	Finish Ch. 3
	3	***Lab 2: Basic Cell Structure***
	4	Review Questions for Quiz
	5	Ch. 3 Quiz
4	1	Begin Ch. 4: Intro. to Cell Basics
	2	Finish Ch. 4 & Review Questions
	3	***Lab 3: Diffusion and Osmosis 1***
	4	Ch. 4 Quiz
	5	Begin Ch. 5: Organelles of the Eukaryotic Cell
5	1	Finish Ch. 5
	2	Review Questions for Quiz
	3	***Lab 4: Diffusion and Osmosis 2***
	4	Ch. 5 Quiz & review for Exam
	5	Unit 2 Exam
6	1	Begin Ch. 6: Basics of Metabolism
	2	Finish Ch. 6
	3	***Lab 5: Enzymes***
	4	Review Questions for Quiz
	5	Ch. 6 Quiz
7	1	Begin Ch. 7: Photosynthesis
	2	Continue Ch. 7
	3	Finish Ch. 7 & Review Questions
	4	Ch. 7 Quiz
	5	Begin Ch. 8: Cellular Respiration

** Occasionally, a lab is scheduled the week before or the week after the textbook reading is assigned. For those weeks, just re-read or review the sections listed in "Preparation."*

WEEK	DAY	LECTURE/LABORATORY
8	1	Continue Ch. 8
	2	Finish Ch. 8 & Review Questions
	3	***Lab 6: The Central Dogma 1***
	4	Ch. 8 Quiz & review for Exam
	5	Unit 3 Exam
9	1	Begin Ch. 9: The Central Dogma
	2	Finish Ch. 9
	3	***Lab 7: The Central Dogma 2***
	4	Review Questions for Quiz
	5	Ch. 9 Quiz
10	1	Begin Ch. 10: The Lac Operon
	2	Finish Ch. 10
	3	***Lab 8: The Lac Operon***
	4	Review Questions for Quiz
	5	Ch. 10 Quiz
11	1	Begin Ch. 11: Recombinant DNA Technology & Genetic Modification
	2	Finish Ch. 11 & Review Questions
	3	***Lab 9: Recombinant DNA Technology***
	4	Ch. 11 Quiz & review for Exam
	5	Unit 4 Exam
12	1	Begin Ch. 12: Mitosis & Cell Division
	2	Finish Ch. 12
	3	***Lab 10: Mitosis & Cell Division***
	4	Review Questions for Quiz
	5	Ch. 12 Quiz)
13	1	Begin Ch. 13: Meiosis
	2	Finish Ch. 13
	3	***Lab 11: Meiosis***
	4	Review Questions for Quiz
	5	Ch. 13 Quiz
14	1	Begin Ch. 14: Basics of Mendelian Genetics
	2	Finish Ch. 14 & Review Questions
	3	***Lab 12: Mendelian Genetics***
	4	Ch. 14 Quiz & review for Exam
	5	Unit 5 Exam

WEEK	DAY	LECTURE/LABORATORY
15	1	Part 1 Review
	2	Part 1 Review
	3	Part 1 Review
	4	Part 1 Review
	5	Part 1 Comprehensive Exam
16	1	Part 2 Intro & begin Ch. 15: Classifying Life
	2	Finish Ch. 15
	3	***Lab 13: Classification***
	4	Review Questions for Quiz
	5	Ch. 15 Quiz
17	1	Begin Ch. 16: Viruses & Prokaryotes
	2	Finish Ch. 16
	3	Review Questions for Quiz
	4	Ch. 16 Quiz
	5	Begin Ch. 17: The Algae: Plant-like Protists
18	1	Finish Ch. 17 & Review Questions
	2	Ch. 17 Quiz
	3	***Lab 14: Kingdom Protista***
	4	Review for Exam
	5	Unit 6 Exam
19	1	Begin Ch. 18: Animal-like and Fungal-like Protists
	2	Finish Ch. 18
	3	***Continue Lab 14: Kingdom Protista****
	4	Review Questions for Quiz
	5	Ch. 18 Quiz
20	1	Begin Ch. 19: Kingdom Fungi
	2	Finish Ch. 19
	3	***Lab 15: Kingdom Fungi***
	4	Review Questions for Quiz
	5	Ch. 19 Quiz
21	1	Begin Ch. 20: Kingdom Animalia
	2	Finish Ch. 20 & Review Questions
	3	Ch. 20 Quiz
	4	Review for Exam
	5	Unit 7 Exam
22	1	Begin Ch. 21: Phylum Porifera
	2	Finish Ch. 21
	3	Review Questions for Quiz
	4	Ch. 21 Quiz
	5	Begin Ch. 22: Phylum Cnidaria
23	1	Finish Ch. 22
	2	Ch. 22 Review Questions
	3	***Lab 16: Phyla Porifera & Cnidaria***
	4	Ch. 22 Quiz
	5	Begin Ch. 23: The Worms

* *Optional: Instead of day 3 lecture continue Lab 14 if you want more time to observe specimens.*

WEEK	DAY	LECTURE/LABORATORY
24	1	Finish Ch. 23 & Review Questions
	2	Ch. 23 Quiz
	3	***Lab 17: The Worms***
	4	Review for Exam
	5	Unit 8 Exam
25	1	Begin Ch. 24: Phylum Mollusca
	2	Finish Ch. 24
	3	***Lab 18: Phylum Mollusca***
	4	Review Questions for Quiz
	5	Ch. 24 Quiz
26	1	Begin Ch. 25: Phylum Arthropoda
	2	Continue Ch. 25
	3	***Lab 19: Phylum Arthropoda***
	4	Continue Ch. 25
	5	Finish Ch. 25
27	1	Review Questions for Quiz
	2	Ch. 25 Quiz
	3	***Lab 20: Phylum Echinodermata***
	4	Begin Ch. 26: Phylum Echinodermata
	5	Finish Ch. 26
28	1	Review Questions for Quiz
	2	Ch. 26 Quiz
	3	***Lab 21: Phylum Chordata 1***
	4	Review for Exam
	5	Unit 9 Exam
29	1	Begin Ch. 27: Phylum Chordata
	2	Finish Ch. 27
	3	***Lab 22: Phylum Chordata 2***
	4	Review Questions for Quiz
	5	Ch. 27 Quiz
30	1	Begin Ch. 28: Kingdom Plantae
	2	Finish Ch. 28 & Review Questions
	3	***Lab 23: Kingdom Plantae 1 (the mosses***
	4	***and ferns)***
	5	Ch. 28 Quiz
31	1	Begin Ch. 29: The Basics of Ecology
	2	Finish Ch. 29 & Review Questions
	3	***Lab 24: Kingdom Plantae 2***
	4	Ch. 29 Quiz & review for Exam
	5	Unit 10 Exam
32	1	Part 2 Review
	2	Part 2 Review
	3	***Lab 25: Ecology***
	4	Part 2 Review
	5	Part 2 Comprehensive Exam

LABORATORY I

THE MICROSCOPE

MATERIALS

- Compound microscope
- Letter 'e' slide
- Glass slides and cover slips
- Clear plastic metric ruler

PREPARATION

In preparation for Lab 1, make sure you have read Chapter 3 "A Short History in Microscopy" in *The Riot and the Dance* (pp. 47–51).

OBJECTIVES

i. Know the parts of the microscope.

ii. Know how to operate the microscope.

i. Know how the microscopic image has been altered compared to the actual object on the slide.

ii. Be able to calculate the total magnification on the microscope.

iii. Be able to calculate the diameter of the field of view at high power if given diameter and magnification values for scan or low power.

EXERCISES

A. Microscope Anatomy

Before you begin looking at stuff through the microscope, it is good to know its parts and what they do so you can operate it intelligently and properly.

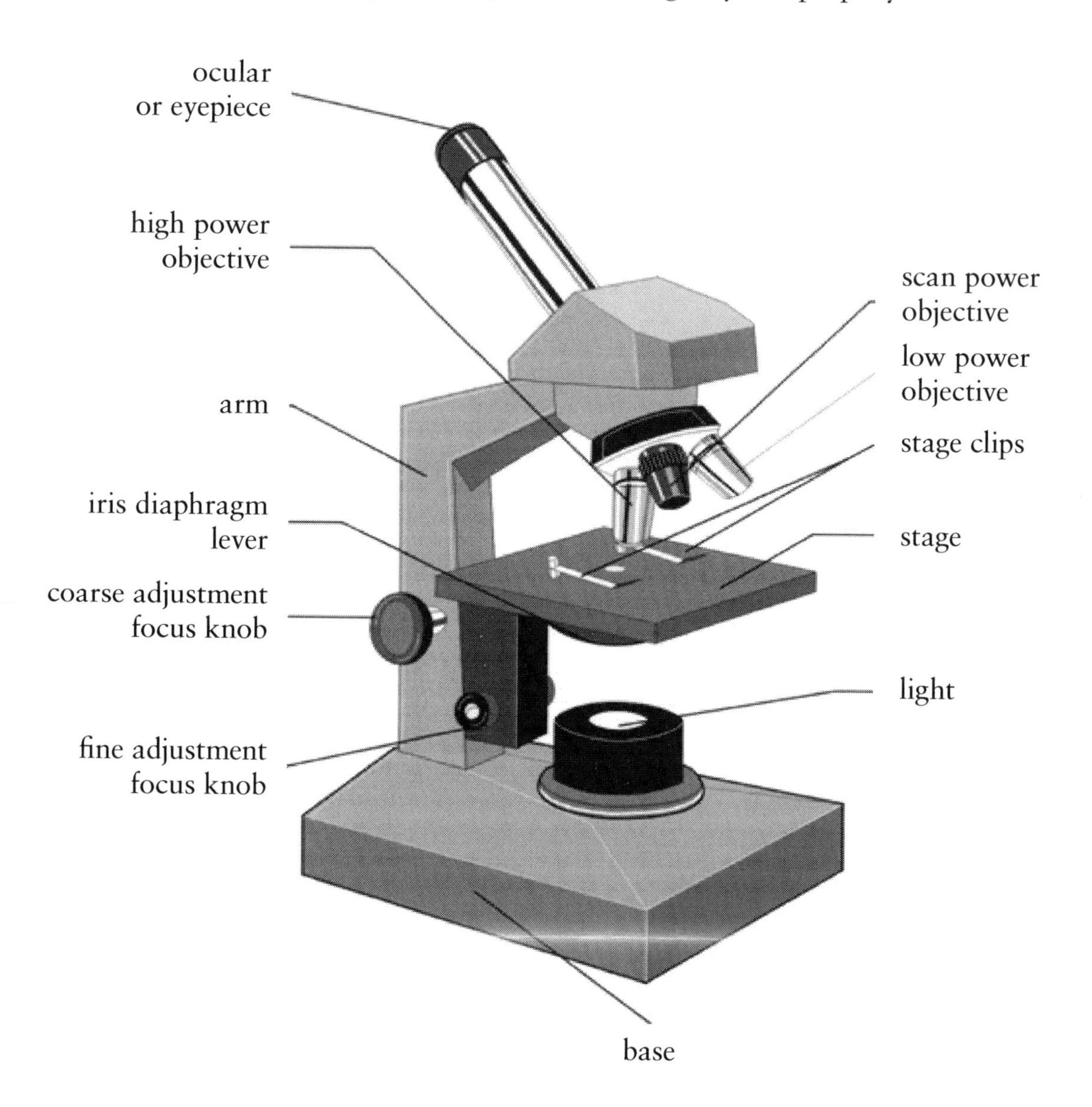

B. The 'e' Slide

1. Get in a comfortable position behind your microscope.
2. Plug in the microscope and turn on the light.
3. Rotate the nosepiece so that the scan objective is in place.
4. Place the 'e' slide on the stage and secure it with stage clips. Looking from the side center the 'e' so that it's positioned directly above the hole in the stage where the light shines through.
5. Peer through the ocular. Using **coarse adjustment focus knob,** bring the 'e' into focus. Use the **fine adjustment focus knob** to bring it into sharp focus if necessary. Adjust lighting as appropriate using the **iris diaphragm lever.**
6. Look at the 'e' on the stage with the naked eye by peering around the side. How has the orientation of the 'e' changed besides being smaller?

   ____

7. Move the slide slightly to the left while watching it through the microscope. Which way did the 'e' move? ______________________

   Move it to the right. Which way did it move? ______________________

   Move the slide away. Which way did it move? ______________________

   Move the slide toward you. Which way did it move? ________________

C. Calculations

1. The total magnification is calculated by multiplying the ocular power by the objective power.

   Total at scan: 10X (ocular) x 4X (scan) = ______________________ (M1)

   Total at low: 10X (ocular) x 10X (low) = ______________________

   Total at high: 10X (ocular) x 40X (high) = ______________________ (M2)

2. To calculate the diameter of the field of view (the circular area you see under the microscope) at high power, follow these steps:

   a. Set the microscope on scan power.

   b. Place a transparent metric ruler on the stage, look through the microscope, and adjust the ruler using your fingers so that you see the metric edge run through the middle of the field of view running left to right.

   c. Focus on the millimeter hash marks along the edge and adjust the ruler so the left edge of a hash mark is at the left border of the field of view.

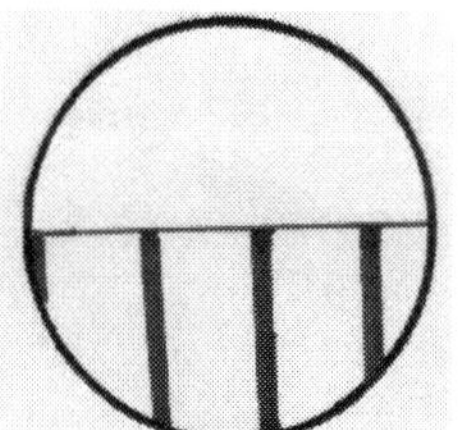

   d. Estimate (to the nearest tenth of a millimeter) the diameter of the field. Record the diameter in millimeters (mm). The above example would measure about 3.5 mm but don't go with this; do it yourself.

   e. Diameter of the field of view at scan power: D1 ______________

   f. Now rotate the nosepiece to low power. Adjust the ruler and record the diameter in mm here.

   g. Diameter of the field of view at low power: ______________

   h. Now rotate the nosepiece to high power. Adjust the ruler and all you'll see is a big black hash mark! You won't be able to measure the diameter accurately. Now what? We calculate that diameter (D2) using simple algebra. Here is the equation:

$$D1 \times M1 = D2 \times M2$$

D1 = diameter of the *field of view* at scan power (measure it)

M1 = magnification at scan power (known)

M2 = magnification at high power (known)

D2 = diameter of the *field of view* at high power (unknown)

3. Rearrange the equation to solve for D2. Divide both sides by M2 and now you can solve for D2. You can also use low power diameter for D1, but then M1 is the total magnification at low power (100X).

   a. $\frac{D1 \times M1}{M2} = \frac{D2 \times \cancel{M2}}{\cancel{M2}}$

   b. What is your estimate of D1? ________

      M1 = 40X

      M2 = 400X

   c. Now solve for D2

      $\frac{D1 \times M1}{M2} = D2 =$ ________ mm

   d. What is D2 in µm? ________ (1 mm is 1,000 µm)

      Example: 0.1 mm is 100 µm & 0.5 mm is 500 µm

LABORATORY 2

BASIC CELL STRUCTURE

MATERIALS

- Glass slides and cover slips
- Methylene blue for staining
- Toothpicks
- *Elodea* (Anacharis) water plant sprig
- The protozoan *Amoeba proteus*
- Your cheek should be available

PREPARATION

In preparation for Lab 2, make sure you have read the sections before diffusion and osmosis in Chapter 4 "Introduction to Cell Basics" in *The Riot and the Dance* (pp. 53–56).

OBJECTIVES

i. Identify the parts of a cheek epithelial cell, Elodea plant cell, and an amoeba which are discernible with the compound light microscope.

ii. Observe, sketch, and label a cheek cell, amoeba, and *Elodea* plant cell.

iii. Know the basic function of pseudopods.

iv. Know what structures plant cells have that human/animal or protozoan cells don't have.

v. Estimate the diameter of a cheek cell, an amoeba, and the length of an *Elodea* cell using the diameter of field of view obtained in Lab 1.

EXERCISES

A. Review: Estimating Cell Size

The diameter is easily calculated by dividing D2 (diameter at high power) by the number of cells (cheek, amoeba, and elodea cells) that could fit across the field of view.

$$\text{Cell diameter} = \frac{\text{diameter of field at current magnification}}{\text{number of cells that fit across the field}}$$

B. Cheek Cells

1. Get a blank slide and firmly scrape the inside of your cheek several times with a toothpick.
2. Rub the 'cheek scrapings' onto the middle of the blank slide.
3. Put one drop of methylene blue on the cheek scrapings.
4. Place the edge of a cover-slip on the edge of the drop and slowly lower it over the drop so that it spreads out evenly under the cover-slip. (This is called a **wet mount**.)
5. Place the slide on the stage and secure it with stage clips. Make sure the scan objective is in position.
6. Center the patch of cheek scrapings so that it's positioned directly above the hole in the stage where the light shines through.
7. Focus as you did before (use coarse adjustment first, then fine), bringing the cheek cells into sharp focus (they should be stained a light blue). Also, adjust lighting. Look for single cheek cells that are separated from clumps

of cells. Pick a nice roundish one and adjust it so that it is in the middle of the field of view.

8. Rotate nosepiece to low power and again center the cell it in the middle of the field of view. Focus using fine adjustment and rotate the nosepiece to high power. The dark blue stained dot in the middle is the **nucleus.** The boundary of the cell is the **cell membrane.** Between these two is the fluid of the cell called **cytoplasm.**

9. Take at least 5 minutes to sketch the cheek cell and label the structures mentioned above. Pay attention to detail.

10. What is the approximate size of the cheek cell you drew?

    __________ mm __________ µm

C. Amoebas

1. Get a blank slide. Using a plastic dropper, suck up a piece of 'skuzzy material' (about a drop) from the bottom of the *Amoeba proteus* jar and place it in the middle of the slide. The drop should have plenty of amoebas in it.

2. Place the edge of a cover-slip on the edge of the drop and slowly lower it over the drop so that the water spreads out evenly under the cover-slip and doesn't trap air bubbles (a wet mount, remember).

3. Place the slide on the stage and secure it with stage clips. Center the patch of 'skuzzy material' so that it's positioned directly above the hole in the stage where the light shines through.

4. View at scan power (always start there). Focus as you did before (always coarse, then fine), bringing the patch of skuzz into focus. Look for an amoeba. There should many to pick from.

5. Once you found an interesting looking one (it should be pretty small on scan power), center it in the middle of the field of view and rotate the nosepiece to low power. Watch it move about using its lobe-like **pseudopods.** Adjust the lighting using the iris diaphragm lever.

6. Using your fingers to move the slide, center the amoeba and rotate the nosepiece to high power. (**Careful: at high power the objective will be very close to the slide so don't use the coarse adjustment, because you could easily bash the lens into the slide**). Using **fine adjustment,** bring it into sharp focus. Adjust lighting using the iris diaphragm lever. Now watch it move for as long as you like. Enjoy. Take note of the grainy **cytoplasm** flowing into pseudopods causing them to extend while noting other pseudopods shrinking as the cytoplasm flows out of them. The flow is called **cytoplasmic streaming.** Some of the larger spheres are **food vacuoles.** These are the result of the amoeba phagocytosing bacteria, other protozoa, or algae. Note the fluid-like **cell membrane** that forms the boundary of the amoeba. You might see a large spherical gray **nucleus** roll along being moved by the cytoplasmic streaming. If you are lucky you'll see a clear spherical area that repeatedly shrinks and enlarges. That is the contractile vacuole (we will discuss that later).

7. Take at least 5 minutes to sketch the amoeba at the top of the next page (even though it's constantly changing shape). Pay attention to detail and don't be in a hurry. Label the structures mentioned above. As it moves, you might have to adjust the slide with your fingers to keep it in view.

8. What is the approximate size of the amoeba you drew?

   __________ mm __________μm

D. Plant Cells

1. Prepare a wet mount of an elodea leaf. (Get a blank slide. Using tweezers pluck a healthy looking leaf off the stem of an elodea plant and place it in the middle of the slide and add one drop of water on top of it. Place the edge of a cover-slip on the edge of the drop and slowly lower it over the drop so that it spreads out evenly under the cover-slip.)
2. Place the slide on the stage and secure it with stage clips. Center the leaf. View it and focus at scan power using coarse then fine adjustment. Switch to low power and bring it into focus (you might move it around to find a nice bright green part of the leaf). Switch to high power and focus using fine adjustment.
3. Note the rows of brick-shaped green cells. The outermost boundary of these plant cells is the **cell wall**. You won't be able to distinguish between the cell wall and the **cell membrane** because the membrane will be pressed up against the inner surface of the cell wall like a full trash bag pressed against the wall of the trash can.
4. Inside these plant cells you will see dozens of oval bright green objects (some may be moving inside the cells). These are the famous organelles called

chloroplasts. They may be more densely packed near the borders of each cell because a **large, central vacuole** occupies the central area of the cell. Chloroplasts are what make plants green and are where the magical process of photosynthesis occurs. Carbon dioxide and water are transformed into glucose (sugar) and oxygen using the energy from sunlight.

5. If you are patient and have keen eyesight you may see a larger light gray **nucleus** among the chloroplasts.
6. Take at least 5 minutes to sketch two or three adjacent plant cells. Again pay attention to detail and don't be in a hurry. Label the structures mentioned above. The vacuole might not be visible but the absence of chloroplasts indicates its presence.
7. What is the approximate size of the plant cell you drew?

   _________ mm _________ µm

LABORATORY 3

DIFFUSION AND OSMOSIS 1

MATERIALS

- A roll of 1 inch wide dialysis tubing
- String
- Plastic droppers
- Metric balance scale or kitchen scale that can weigh ounces
- One bottle of corn syrup (Karo)
- Three bowls that can hold up to 2 cups

PREPARATION

In preparation for Labs 3 and 4, make sure you have read the sections on diffusion, osmosis, and tonicity in Chapter 4 "Introduction to Cell Basics" in *The Riot and the Dance* (pp. 56–63).

Prepare the various solutions and dialysis tubing in advance.

1. Cut three lengths of dialysis tubing, each about four inches long. Soak them in water for several hours (overnight is great).
2. Cut six lengths of string about two inches long.
3. Prepare 1 cup of 50% Karo syrup (½ cup water; ½ cup Karo). Mix it well.

4. Prepare 2 cups of 25% Karo syrup (1½ cup water; ½ cup Karo). Mix it well.
5. Tightly tie off the end of the three lengths of dialysis tubing with the string (double knot).
6. Have three labeled bowls containing the appropriate solution:
   a. Hypertonic: 1 cup of 50% Karo syrup
   b. Hypotonic: 1 cup of water (0% Karo syrup)
   c. Isotonic: 1 cup of 25% Karo syrup

OBJECTIVES

i. Know the definitions of solute, solvent, solution, diffusion, osmosis, and Brownian movement.
ii. Describe the effect of heat and concentration on the rate of diffusion.
iii. Define tonicity and the three types of tonicity: hypertonic, hypotonic, and isotonic.
iv. Given the tonicity, predict the direction of water movement through a semi-permeable membrane.
v. Define turgidity, plasmolysis, and cytolysis.
vi. Explain the function of the central vacuole of plant cells.

EXERCISES

A. Definitions

Use the textbook glossary to define the following words, and think about how these definitions apply while doing the experiment.

solute: ____

solvent: ____

solution: ____

diffusion: ______________________________

equilibrium: ______________________________

osmosis: ______________________________

Brownian motion: ______________________________

tonicity: ______________________________

hypertonicity: ______________________________

hypotonicity: ______________________________

isotonicity: ______________________________

B. Osmosis in Simulated Cells

Simulated "Cell" in a Hypertonic Solution

1. Using the dropper, fill one dialysis tubing with 25% Karo syrup. Fill it so that there is enough empty tubing to tightly tie off the other end of the bag with a piece of string.
2. Lightly blot the outside of the bag to remove excess syrup and weigh it in grams or ounces. In the table on p. 16, record the weight in the top box of the **hypertonic** column.
3. Place this syrup bag in the dish containing 50% syrup (hypertonic solution). This represents a cell in a hypertonic solution, i.e. an environment where the solute concentration is higher outside the cell than inside the cell.

Simulated "Cell" in a Hypotonic Solution

4. Do the same thing to the second dialysis tubing (steps 1 & 2 in the previous section, except record its weight in the top box of the hypotonic column on p. 16). Don't worry if they have different amounts of syrup in them—change in weight is what matters, not final weight.

5. Place this syrup bag in the dish containing just water (0% syrup). This represents a cell in a hypotonic solution, i.e. an environment where the solute concentration is lower outside the cell than inside the cell.

Simulated "Cell" in an Isotonic Environment

6. Do the same thing to the third dialysis tubing, except record its weight in the top box of the isotonic column below.
7. Place this syrup bag in the dish containing 25% syrup (isotonic solution).
8. Now let all three syrup bags soak in their three different environments for at least 30 minutes.
9. Based on what you know of osmosis and considering the fact that water can diffuse across the dialysis tubing but syrup cannot, predict what the results will be on the final weight of the bags?

   Which tube do you think will gain water? ____________________

   Which tube do you think will lose water? ____________________

   Which tube should stay approximately the same? ______________
10. After 30+ minutes, blot dry each bag, reweigh, and record its final weight in the appropriate column below. Through subtraction, determine how much weight each bag lost, gained, or did neither.

	Hypertonic (50%)	Hypotonic (0%)	Isotonic (25%)
Initial weight			
Final weight			
Weight change			

Were you correct in your predictions? ______________________________

Explain why the bag in the hypertonic solution lost weight.

____

____

____

____

Explain why the bag in the hypotonic solution gained weight.

____

____

____

____

Explain why the bag in the isotonic solution stayed about the same.

____

____

____

____

If your results were different than what they were supposed to be, can you explain what may have gone wrong in setting up the experiment?

____

____

____

____

LABORATORY 4

DIFFUSION AND OSMOSIS 2

MATERIALS

- 100 yellow and 30 red pop beads
- Tea kettle to boil water
- 4 tall clear glasses
- 2 tea bags
- 1 tablespoon of table salt
- Two fresh crisp celery stalks (both 3 inches long and roughly equal in width)
- Microscope and 1 slide
- Fresh Elodea sprigs soaking in fresh water
- A small container with a very flat bottom such as a casserole dish or shoebox lid

PREPARATION

- The night before, fill two glasses with one cup of cool tap water in each.
- Add about a tablespoon of table salt to one of the glasses.
- Place one 3-inch celery stalk in each glass, and let them soak overnight.

OBJECTIVES

i. Describe the effect of heat and concentration on the rate of diffusion.
ii. Define turgidity, plasmolysis, and cytolysis.
iii. Predict the direction of water movement when celery is exposed to a hypertonic and hypotonic solution.
iv. Predict the direction of water movement when elodea is exposed to a hypertonic solution.
v. Explain the function of the central vacuole of plant cells.

EXERCISES

A. Definitions

Use the textbook glossary to define the following terms, and think about the definitions while doing the experiment.

turgidity: ______________________________

plasmolysis: ______________________________

cytolysis: ______________________________

B. Simulation of Brownian Motion and Diffusion Using Pop Beads

1. Set the shallow, flat-bottomed container on a flat surface, and spread 100 yellow pop beads evenly across the bottom.
2. Place 30 red pop beads in a cluster on one end surrounded by the yellow. Note that the opposite end of the container has a low concentration of red beads.
3. The red beads represent molecules of a solute placed in a solvent.
4. While **keeping it on a flat surface**, rapidly shake the container from side to side in all horizontal directions. This represents Brownian motion. Watch

the red beads dance randomly about. In their random movements they will move from an area of high concentration of red beads to an area of low concentration of red beads. **This represents diffusion.** Eventually the red beads will be evenly distributed throughout the yellow beads. At that point it has reached **equilibrium.**

C. Diffusion of Tea

1. Fill two glasses with water, one with very cold or ice cold water and the other with boiling water.
2. Next, at the same time, place one tea bag in the ice water and the other in the boiling water. (Do not swirl the tea bags around.)
3. Note diffusion rates at the different temperatures.

   Which water temperature caused the tea to diffuse faster? ____________

   Describe why this happened in terms of Brownian motion.

   ____

D. Osmosis in Celery (Macroscopic) in a Hypotonic and Hypertonic Environment

This demonstration must be prepared the night before.

Based on what you know about osmosis, which way do you think water will move in each stalk of celery?

Salt soaked: (circle one)

a. Into the celery cells
b. Out of the celery cells

Fresh water soaked: (circle one)

a. Into the celery cells
b. Out of the celery cells

Pick up the celery stalks. Bend them and feel their texture.

Which one feels more bendable and rubbery (it experienced plasmolysis and a loss of turgor pressure)? ____________________

Which one feels firm and crisp (it maintained high turgor pressure)? ________

Why didn't the crisp stalk experience cytolysis?

____

What environment was the crisp celery stalk kept in?

a. Hypotonic
b. Hypertonic

What environment was the rubbery celery stalk kept in?

a. Hypotonic
b. Hypertonic

Describe what happened in terms of osmosis to the cells in the celery stalks. __

____

____

Were you correct in your prediction? ____________________

E. Elodea (Microscopic) in a Hypotonic and Hypertonic Environment

1. Prepare a wet mount of an elodea leaf (same as in lab 2).
2. Place the slide on the microscope and observe the leaf cells under scan, low, and high power.

   Predict what will happen to the leaf cells when they experience a salty (hypertonic) environment. ____________________

   ____

   ____

3. Using a dropper carefully place a small drop of salt water (using the salt water glass from the celery experiment) on the **left edge** of the cover slip

while placing a small square of paper towel on the **right edge** so that it touches the water under the cover slip.

4. Let the paper towel wick the water out from under the cover slip. This should simultaneously pull the salt water under the cover slip.
5. While doing this, have the plant cells in focus at high power. Try to keep them in focus (adding and removing water can cause the leaf to go out of focus). As the salt water is wicked under the cover slip it comes in contact with the leaf. Watch carefully what happens to the plant cells. If nothing happens, add another drop of salt water and wick some more with a piece of paper towel as before.

   Describe what happens to the cells over the next few minutes.

   ____

   ____

   ____

   ____

   Did the cell membrane pull away from the cell wall?________________

   What happened to the chloroplasts?______________________________

   What is happening to these plant cells?

   a. Plasmolysis
   b. Cytolysis
   c. Gain in turgor pressure

   Did something like this happen to the salt-soaked celery?_____________

LABORATORY 5

ENZYMES

MATERIALS

- Water
- 1 potato
- Blender
- Wire mesh strainer
- A few crystals of copper sulfate ($CuSO_4$)
- 1 bottle of hydrogen peroxide
- 10 test tubes
- Metric ruler
- Test tube holder
- A few drops of distilled white vinegar
- A graduated dropper (that can at least dispense about 3 ml)
- Stopwatch

PREPARATION

In preparation for Lab 5 make sure you have read Chapter 6 "Basics of Metabolism" in *The Riot and the Dance* (pp. 85–93).

Catalase is an enzyme that is very common in the cells of plants and animals. A common waste product of metabolism is hydrogen peroxide (H_2O_2) and is toxic to the cell. Therefore God has provided the cell with an enzyme that can convert hydrogen peroxide into water (H_2O) and oxygen gas (O_2), which are both harmless. One catalase enzyme can break down millions of hydrogen peroxide molecules per minute! When you mix potato juice (which contains catalase) with hydrogen peroxide it will begin to bubble (the bubbles are oxygen gas, a product of the reaction) which tells you that catalase is doing its job. Water (the other product) produced by the reaction will not be measurable or noticeable.

- Make filtered potato juice the day before the lab:
    1. Peel one potato and cut it up into 3 or 4 pieces.
    2. Place it in a blender with 1 cup of water.
    3. Blend at fastest setting (like you're making a potato smoothie)
    4. Filter the blended potato through a screen strainer to remove the potato pulp. You should have a little more than a cup of pinkish-brown potato juice for your experiments.
    5. Refrigerate and let the potato juice sit overnight.
- Make copper sulfate ($CuSO_4$) solution: Dissolve a few crystals in 2 ml of water. Copper sulfate doesn't dissolve easily, so don't try to get all the crystals to dissolve. If you stir the water until it turns bluish, that's fine.

OBJECTIVES

i. Define enzyme, active site, substrate, denaturation, and non-competitive (allosteric) inhibition.

ii. Know the catalase chemical reaction.

(substrate) (enzyme) (product) (product)

$$2H_2O_2 \rightarrow \text{catalase} \rightarrow 2H_2O + O_2$$

peroxide water oxygen gas

iii. Know how to prepare potato juice (which contains plenty of catalase enzyme).

iv. Describe the effect of enzyme concentration, temperature, low pH, and a heavy metal on enzyme activity.

EXERCISES

A. Definitions

Use the textbook glossary to define these terms. Again, think about the definitions while doing the experiment.

enzyme: ____________________

active site: ____________________

substrate: ____________________

denaturation: ____________________

non-competitive (allosteric) inhibition: ____________________

B. Experiments

Catalase reaction at room temperature

1. Add 3 ml of potato juice to a test tube.
2. Add 3 ml of hydrogen peroxide to another test tube.
3. Pour the hydrogen peroxide into the potato juice tube. Cap with thumb, flip upside down once to mix. The reaction will start. The visible product of the reaction will be oxygen bubbles.
4. Start stopwatch immediately after mixing. Let the reaction go for 1 minute.

5. At the end of 1 minute measure the height of the column of bubbles (from the surface of the liquid to the top of the bubble column). Record the height in mm in the table on p. 29 in the "control" column.

Catalase reaction on ice (the ice bath is to slow down Brownian motion of the enzyme and the substrate)

6. Add 3 ml of potato juice to a test tube (chill in ice bath)
7. Add 3 ml of hydrogen peroxide to another test tube (chill in ice bath)
8. Pour the hydrogen peroxide into the potato juice test tube. Cap with thumb, flip upside down once to mix. The reaction will start.
9. Place the mixture back into the ice bath.
10. Start stopwatch immediately after mixing. Let the reaction go for 1 minute.
11. At the end of 1 minute measure the height of the column of bubbles. Record the height in mm in the "On Ice" column of the table.

Catalase reaction after boiling the potato juice

12. Add 3 ml of potato juice to a test tube.
13. With test tube holder, hold potato juice test tube in boiling water for 30 seconds.
14. Add 3 ml of hydrogen peroxide to another test tube.
15. Pour the hydrogen peroxide into the potato juice test tube. Cap with thumb, flip upside down once to mix.
16. Start stopwatch immediately after mixing. Let the reaction go for 1 minute.
17. At the end of 1 minute measure the height of the column of bubbles. Record the height in mm in the "Boiled Enzyme" column of the table.

Catalase reaction at low pH

18. Add 3 ml of potato juice to a test tube.
19. Add a couple drops of distilled white vinegar to the potato juice test tube (mix it well).
20. Add 3 ml of hydrogen peroxide to another test tube.
21. Pour the hydrogen peroxide into the potato juice test tube. Cap with thumb, flip upside down once to mix.
22. Start stopwatch immediately after mixing. Let the reaction go for 1 minute.
23. At the end of 1 minute measure the height of the column of bubbles. Record the height in mm in the table below in the "Low pH" column.

Catalase reaction with heavy metal ($CuSO_4$)

24. Add 3 ml of potato juice to a test tube.
25. Add 3 or 4 drops of copper sulfate ($CuSO_4$) solution to the potato juice test tube (mix it well)
26. Add 3 ml of hydrogen peroxide to another test tube.
27. Pour the hydrogen peroxide into the potato juice test tube. Cap with thumb, flip upside down once to mix.
28. Start stopwatch immediately after mixing. Let the reaction go for 1 minute.
29. At the end of 1 minute measure the height of the column of bubbles. Record the height in mm in the table below in the "Heavy Metal" column.

	Control (room temp)	On Ice (~32⁰ F)	Boiled Enzyme	Low pH	Heavy Metal
Height of O_2 bubble column					

Under what conditions did the catalase have maximum performance?

___

Why was there less product made when both the enzyme and substrate were kept cold on ice? ______________________________

____

Why were there very few or no oxygen bubbles at the low pH?

____

Why were there very few or no oxygen bubbles when a heavy metal was present? ______________________________

____

LABORATORY 6

THE CENTRAL DOGMA 1

MATERIALS

- A computer to watch online videos

PREPARATION

Make sure you have read the "DNA Replication" section in Chapter 12 "Mitosis and Cell Division" and the "RNA Transcription" section in Chapter 9 "The Central Dogma" in *The Riot and the Dance* (pp. 121–126; 143–145).

OBJECTIVES

i. Be able to describe the process of DNA Replication. Given a short sequence of DNA, be able to produce a complimentary strand of DNA.
ii. Be able to describe the process of RNA Transcription. Given a short sequence of DNA, be able to produce a complimentary strand of RNA.
iii. Know the definitions of the terms listed in the exercises.

EXERCISES

A. Definitions

Use the textbook glossary to define these terms, and think about the definitions while doing the exercises.

gene: ______________________________

nucleotide: ______________________________

DNA replication: ______________________________

DNA polymerase: ______________________________

RNA transcription: ______________________________

RNA polymerase: ______________________________

B. Video

DNA Replication

1. Visit http://logospressonline.com/content/RiotLinks.pdf (you'll want to bookmark that document in your browser for easy reference in future labs). Click the link for the Laboratory 6 video, "Central Dogma," and watch up to the 6:50 mark. Maybe watch it twice to get the idea down. Once you're ready, proceed with the lab.

Example of DNA Replication

2. Using this short double-stranded segment of DNA, I'll show you how DNA replication constructs the new strands from the old strands.

T G T T C G A G C G T C A T T T C A A C C

A C A A G C T C G C A G T A A A G T T G G

3. First, I unzip the double-stranded DNA (above) into two single strands, like so. What enzyme unzips the strands apart?

____

TGTTCGAGCGTCATTTCAACC

ACAAGCTCGCAGTAAAGTTGG

4. Second, using the upper strand as a template, I put in the complimentary bases (below the strand) in ***bold italic type*** like so...

TGTTCGAGCGTCATTTCAACC
ACAAGCTCGCAGTAAAGTTGG

...and with the lower strand, I put in (above the strand) the complimentary bases in ***bold italic type*** like so.

TGTTCGAGCGTCATTTCAACC
ACAAGCTCGCAGTAAAGTTGG

The unbolded strands are the old parent strands and the **bolded** strands are the new daughter strands. Note that these two double strands above are identical to each other. The difference is that the new strand in the upper is identical to the old strand in the lower, and vice versa.

What is the enzyme that inserts complimentary nucleotides so they base pair with the parental strands to build the new daughter strands?

____

5. Now you do the same with this sequence of DNA. You can use underlining rather than bolding to differentiate the new strands.

GTAATGTGAATTGCAGAATTCAGTGAA
CATTACACTTAACGTCTTAAGTCACTT

C. RNA Transcription

The upper strand is the beginning of a gene for the protein hormone called insulin.

CCATAGCACGTTACAACGTGAAGGTAA
GGTATCGTGCAATGTTGCACTTCCATT

Below, the gene (DNA) is unzipped.

C C A T A G C A C G T T A C A A C G T G A A G G T A A

G G T A T C G T G C A A T G T T G C A C T T C C A T T

6. Now that they are separated use a pen to transcribe the upper strand. That means make an mRNA strand by putting in RNA bases that are complimentary to each DNA base. Ignore the lower strand because it isn't part of the gene. Remember that RNA does not have T (thymine). Instead it has U (uracil). If DNA has an A then put in a U for the mRNA strand. However, if DNA has a T put in an A for the RNA strand. The GC base pairing is the same. Be meticulously careful. You don't want to create a mutation.

7. The mRNA strand you made should be:

   G G U A U C G U G C A A U G U U G C A C U U C C A U U

   If your mRNA doesn't match, find where you went wrong.

   a. Where does RNA transcription occur in the cell?

   ____

   b. What enzyme transcribes the DNA?

   ____

LABORATORY 7

THE CENTRAL DOGMA 2

MATERIALS

- A computer to watch online videos.

PREPARATION

Make sure you have read the "Protein Translation" section in Chapter 9 "The Central Dogma" in *The Riot and the Dance* (pp. 124–128). Have a spacious table surface available and have all the parts available from the Protein Translation kit.

OBJECTIVES

i. Be able to divide the mRNA up into codons.
ii. Be able to describe in your own words, the process of Protein Translation.
iii. Using a genetic code table, determine the amino acid sequence from a segment of DNA.
iv. Know the names and functions of the 3 types of RNA.
v. Know the definitions of the words listed in the exercises.

EXERCISES

A. Definitions

Use the textbook glossary to define these terms, and think about the definitions while doing the exercises.

protein translation: ______________________________

ribosome: ______________________________

mRNA: ______________________________

codon: ______________________________

tRNA: ______________________________

anticodon: ______________________________

mRNA binding site: ______________________________

P–site: ______________________________

A–site: ______________________________

peptide bond: ______________________________

B. First Video Demonstration

1. Visit http://logospressonline.com/content/RiotLinks.pdf. Click the link for the Laboratory 7 video, which is the "Central Dogma" video continued, and watch up from the 6:45 mark to 23:20 (the end). If needed, watch the segment twice to get the idea down. Then proceed with the lab.

C. The Genetic Code

1. Group the mRNA codons (from Lab 6) by putting a short vertical line between groups of three bases on the following mRNA transcript (the one made in your transcription exercise). These groups of three are called codons. The lines serve to separate each codon clearly.

   G G U A U C G U G C A A U G U U G C A C U U C C A U U

   Like so...

   G G U|A U C|...

2. Use this Genetic Code table to decipher each codon. Translate these nine codons into the first nine amino acids of the insulin protein.

FIRST BASE	SECOND BASE				THIRD BASE
	U	C	A	G	
U	Phe	Ser	Tyr	Cys	U
	Phe	Ser	Tyr	Cys	C
	Leu	Ser	STOP	STOP	A
	Leu	Ser	STOP	Trp	G
C	Leu	Pro	His	Arg	U
	Leu	Pro	His	Arg	C
	Leu	Pro	Gln	Arg	A
	Leu	Pro	Gln	Arg	G
A	Ile	Thr	Asn	Ser	U
	Ile	Thr	Asn	Ser	C
	Ile	Thr	Lys	Arg	A
	Met	Thr	Lys	Arg	G
G	Val	Ala	Asp	Gly	U
	Val	Ala	Asp	Gly	C
	Val	Ala	Glu	Gly	A
	Val	Ala	Glu	Gly	G

Let's decipher the first codon, GGU. The first base G lines up with the four bottom rows (bottom sixteen boxes) of the Genetic Code table. The second base narrows our choices down. The second base G is the last column. Look where the last column intersects with the bottom four rows. That covers four boxes, all of which have Gly, which stands for the amino acid glycine. The third base U tells you which of the four boxes to select. In this case it doesn't matter because the third base could be U, C, A, or G and it would still code for glycine. There is redundancy in the code (several different codons can code for a particular amino acid) because there are 20 different amino acids but 64 different codons.

Where does protein translation occur?________________________________

What amino acid does the codon *AUG* code for?______________________

What amino acid does the codon *CAU* code for?______________________

What information does the mRNA contain? ___________________________

____

What type of bond forms between two amino acids while positioned in the P and A site of the ribosome? _______________________________________

What is the function of tRNA?_______________________________________

What part of the tRNA molecule hooks to the codon of the mRNA molecule during translation? ____

What is the function of the ribosome?_________________________________

____

What two types of biomolecules are ribosomes made of?

a. ______________________________

b. ______________________________

LABORATORY 8

THE LAC OPERON

MATERIALS

- A computer to watch an online video
- A 6-foot light-colored extension cord (this will get marked up, but not ruined)
- A black Sharpie pen
- 6 12-inch pipe cleaners—3 white and three other colors
- A 1¼-inch binder clip
- A white ink marker (or a sticker or masking tape for a label)
- 1 Fisher-Price Snap-Lock bead
- Double-sided tape.
- 20 pop beads
- A wiffle ball
- Scissors

PREPARATION

Make sure you have read Chapter 10 "The Lac Operon" in *The Riot and the Dance* (pp. 131–133).

OBJECTIVES

i. Be able to describe the overall process, the logic, and the goal of an operon.

ii. Know how to describe the overall process and the goal of this operon while acting it out using the materials listed above, including all the components and their functions.

EXERCISES

A. Definitions

Use the textbook glossary to define the following terms, and think about the definitions while doing this exercise.

operon: ____________________

repressor: ____________________

operator: ____________________

promoter: ____________________

lactose (inducer): ____________________

β-galactosidase (lactase): ____________________

permease: ____________________

B. Protein Translation Video Demonstration

1. Visit http://logospressonline.com/content/RiotLinks.pdf. Click the link for the Laboratory 8 video, "Lac Operon," and watch the whole thing (to 16:37).

C. Act Out the Lac Operon

Make a crude model of a bacterial chromosome in a single E. coli cell.

1. The room you're in represents the cell.

2. Plug the ends of the extension cord together to make a big loop. The cord represents the bacterial chromosome (DNA containing the genes of the *E. coli* cell).
3. Mark off a three-inch segment of cord with a sharpie and label it "Promoter." Adjacent and to the right of the promoter, mark off a three inch segment of cord and label it "Operator." Adjacent and to the right of the operator, mark off about 12 inches of cord and label it "Gene for β-galactosidase." Immediately next to that mark off 12 inches of cord and label it "Gene for Permease." Immediately next to that mark off about 12 inches of cord and label it "Gene for Transacetylase."
4. Label the binder clip with a white ink marker (or use a white label and your black Sharpie), "Repressor Protein," and clip it to the "Operator" segment of cord.
5. Using a sturdy pair of scissors, carefully cut a slot lengthwise on one side of the Snap-Lock bead. The slot should be about ¼-inch wide and fit snuggly on the extension cord. Label it with the Sharpie, "RNA Polymerase."
6. Stick a short length of double-sided tape on the back side of the binder clip.
7. Join the pop beads into 10 pairs representing lactose molecules. Remember lactose is a disaccharide of glucose and galactose. It is also called the "Inducer."
8. About two feet to the left of the "Promoter" mark off about 12 inches of cord and label it "Gene for Repressor."
9. The three white pipe cleaners are models of the mRNA strands made from the transcription of the three genes mentioned in #3.
10. The three colored pipe cleaners represent the proteins "β-galactosidase," "Permease," and "Transacetylase."
11. The wiffle ball represents a ribosome.

12. *E. coli* absorbs and breaks down mostly the monosaccharide glucose for its energy needs, but has the ability to make some enzymes that help absorb and break down the disaccharide lactose. Since lactose isn't its usual food source, it makes sense that *E. coli* does not make enzymes specific for lactose metabolism if lactose isn't available. *E. coli* does this by producing a repressor protein. Note that the "Repressor protein" (the binder clip) is latched on to the "Operator" segment of DNA. This repressor protein acts as a 'road block' to the "RNA polymerase."
13. Take the Snap-Lock bead labeled "RNA polymerase," bind it to the "Promoter," and move it to the right in order to transcribe the genes. Since the "Repressor" is blocking the RNA polymerase at the "Operator" region, it can't transcribe anything. This is why we say, "These genes are turned off."
14. Now let's say glucose is no longer available for *E. coli's* food, but lactose is available. *E. coli* now needs to make the enzymes, β-galactosidase and permease if it is to take advantage of this available food source.
15. Enter the "Inducer" lactose. Take one of the lactose models (two small pop-it bead units) and have it come into the cell (along with 3 or 4 others). Have one stick to the two-sided tape on the "Repressor protein." Binding it to the Repressor causes the Repressor to change shape. Demonstrate this by unclipping the "Repressor protein" from the "Operator" region.
16. Note what happened. The "road-blocking" repressor is now no longer in the way of RNA polymerase. Now we can say, "These genes are turned on."
17. Have the RNA polymerase bind to the "Promoter" by slipping the extension cord into its slot. Then freely slide the RNA polymerase down the "Operator" and over the "Genes for β-galactosidase and permease. As you slide it over the genes, have a foot long piece of white pipe cleaner (representing the mRNA of β-galactosidase) peel off of its corresponding

gene as if it was a freshly made mRNA coding for β-galactosidase. Do the same with the permease and transacetylase genes.

18. Translation: Have the three white pipe cleaners (mRNAs) go to the whiffle ball (ribosome). Thread the first white pipe cleaner (β-galactosidase mRNA) through a hole in the whiffle ball and simultaneously have a colored pipe cleaner peel off the wiffle ball as the white one slides through. The colored pipe cleaner represents an unfolded β-galactosidase protein (enzyme) freshly translated at the ribosome. Fold up the β-galactosidase protein into a ball-like shape. It is now ready to do its job; breaking down lactose. Do the same with the permease and the transacetylase. Fold these two up into a ball as well.
19. Have the β-galactosidase enzyme start doing its job; grabbing the lactose (two pop-it bead units) and breaking them apart. This is beginning the process of digesting them down as a food source.
20. Meanwhile have the permease protein go to the wall of the room and tape it there. This represents permease embedding itself in the cell membrane to form a gateway to help more lactose to enter the cell. Have the other 5 or so enter the cell through the pretend gateway. As these enter, have the β-galactosidase enzyme break these in two as well. You probably need to pop them apart with your fingers but pretend the pipe cleaner ball (β-galactosidase) did it.
21. Lastly have β-galactosidase break down the lactose stuck to the "Repressor protein" removing it from the Repressor. When it does, clip the "Repressor protein" back on the "Operator" region. Now it is back in the blocking position. We can now say, "These genes are turned off." The purpose of the lac operon is to only make proteins involved in lactose utilization if lactose is available.

22. If you have time, act out the whole process again and describe what's happening as you go. If there isn't time to act it out again, review how it was acted it out in your mind while it is still fresh.

LABORATORY 9

RECOMBINANT DNA TECHNOLOGY

MATERIALS

- 100 yellow and 100 red pop beads
- A shoebox—preferably plain so you're not distracted by the Nike, Adidas, etc., emblazoned on the box. (A logo makes it harder to pretend the box is a bacterial cell.)
- Black Sharpie
- Scissors

PREPARATION

Make sure you have read Chapter 11 "Recombinant DNA Technology" in *The Riot and the Dance* (pp. 135–141).

OBJECTIVES

i. Know the main tools in recombinant DNA technology.

ii. Know the steps used in cloning a gene and transforming a recipient cell with a recombinant plasmid (you will act out this process):

- Purify DNA containing gene of interest

- Cut out gene of interest with appropriate restriction enzyme(s)
- Separate gene of interest from DNA it was cut from
- Open up purified plasmid with same restriction enzyme so it will have matching sticking ends
- Add gene of interest to opened up plasmid
- Splice gene of interest into plasmid using DNA ligase thus creating recombinant DNA
- Transformation of recipient cell with recombinant DNA (plasmid)

EXERCISES

A. Definitions:

Use the textbook glossary to define the following terms, and think about the definitions while doing this exercise.

plasmid: ______________________________

restriction enzyme: ______________________________

sticky ends:______________________________

DNA ligase: ______________________________

transformation: ______________________________

conjugation: ______________________________

transduction: ______________________________

B. Acting out the procedure of recombinant DNA technology

1. Using a length of approximately 100 red pop beads, pop off segments of about 40 beads on each end. This represents cutting a piece of DNA (containing a gene of interest) out of a certain genome (using a restriction enzyme). This middle piece will be about 20 beads long. Suppose that this red segment of DNA is the gene for human insulin. (In a laboratory, this

would be done in a test tube containing DNA having the desired gene. After cutting with the enzyme, the desired gene would be separated from the rest of the DNA using a technique called electrophoresis.)

2. Have the yellow strand of 100 beads in a hoop like a necklace. This represents a purified bacterial plasmid (in a test tube, not in a cell). Act like a restriction enzyme by popping the yellow hoop open in one place. This represents the cutting of a plasmid with the same restriction enzyme used in step 1. The same restriction enzyme ensures that all the cut ends will be "sticky ends" and will bond to each other. (In a laboratory, this would be done in a test tube containing billions of copies of the exact same plasmid. A tiny amount of a certain type of restriction enzyme is added to the test tube containing the plasmid.)
3. Place the red segment (the insulin gene) from step 1 with the opened-up yellow plasmid from step 2. Place the two ends of the red segment and line them up with the two ends of the opened up plasmid. Add DNA ligase (your hands act like DNA ligase by popping the red segment into the yellow plasmid). Now you have recombinant DNA in a test tube.

C. Transformation

This involves getting the recombinant DNA into an actual cell.

1. Label the shoebox with the Sharpie: "Bacterial cell." Carefully cut or poke small holes in the box with scissors. (This perforated box represents a bacterial cell that was treated with certain chemicals that make it receptive to taking in foreign DNA.) In a laboratory, billions of cells in a test tube are treated this way, but we only need one pretend cell to get the idea across.
2. Place the yellow/red hoop (recombinant DNA) on the box (cell surface) and push it into the box through one of the holes. **This is transformation.** (In a laboratory, this is done by adding the recombinant DNA from its test tube

to another test tube containing a suspension of chemically-treated bacterial cells. Once the recombinant DNA and cell suspension are thoroughly mixed, the test tube is placed under a stream of hot tap water for a short time. The hot water just flows around the tube and heats up the contents; it doesn't go into the tube. This heat shock causes some of the cells to 'slurp up' the external DNA. The cells are then put into a nice, nutrient-rich cushy environment to recover from all this chemical abuse).

3. However, how do we find out which cells got transformed? Maybe only a few cells got transformed out of several billion chemically treated cells. Here's the cool part. The recombinant DNA plasmid also has another gene on it coding for antibiotic resistance. So to find out which cells were transformed, the cell suspension is smeared out onto a Petri-dish that has a particular antibiotic in it (the same antibiotic that the plasmid has resistance to). You may have smeared billions of cells on the Petri-dish but **all the cells that did not get transformed, die in the presence of the antibiotic.** Those few cells that got transformed have the plasmid that gives them antibiotic resistance. So let's say 20 cells were transformed out of several billion. Each one of those cells grows into a colony on the Petri-dish. Every cell in each colony has the plasmid because during cell division all the DNA (including the plasmid) is replicated. Now all the surviving descendant cells have the recombinant plasmid too and consequently are resistant to the antibiotic.

4. Now we have genetically engineered bacteria with a plasmid containing the human insulin gene. The bacteria can now be cultured. They busily transcribe and translate the insulin gene and manufacture large amounts of insulin. Techniques have been developed to extract this precious insulin for those who suffer from Type I diabetes. Since it is human insulin, it doesn't cause unpleasant allergic reactions that animal-derived insulin can cause.

LABORATORY 10

MITOSIS AND CELL DIVISION

MATERIALS

- A computer to watch online videos.
- Compound microscope
- Onion (*Allium*) root tip slide
- 12 thin skeins of embroidery floss: 2 light blue, 2 dark blue, 2 pink, 2 red, 2 light green, 2 dark green (Hang on to these, because you'll need them again for Lab 11.)
- Double-sided tape
- Two small, clear wastepaper bags

PREPARATION

Make sure you have read "Interphase" and "Mitosis" sections in Chapter 12 "Mitosis and Cell Division" in *The Riot and the Dance* (pp. 145–151).

OBJECTIVES

i. Know the definitions of interphase, mitosis, and cytokinesis.

ii. Be able to identify each phase of mitosis and cytokinesis in *Allium* root tips.

iii. Be able to define and describe the process of interphase, the phases within it (G1, S, and G2), and the reasons for those phases.

iv. Be able to describe the process of mitosis, the phases within it (prophase, metaphase, anaphase, and telophase) and the reasons for those phases. The textbook gives the reasons for each phase.

v. Be able to describe the animal and plant versions of cytokinesis (cleavage furrowing and cell plate formation).

EXERCISES

A. Definitions

interphase (include G1, S, G2): ____

____

____

____

____

____

mitosis (include prophase, metaphase, anaphase, and telophase):

____

____

____

____

____

____

____

____

____

cytokinesis: ____

____

B. Video demonstration

Visit http://logospressonline.com/content/RiotLinks.pdf. Click the link for the Laboratory 10 video, "Mitosis," and watch the whole thing (to 12:22). This is what you'll be acting out in Section D. Nothing helps memorize a concept as much as acting it out.

C. Examination of cells undergoing mitosis

1. The onion root tip is a very fast growing area in the plant. When the slide was made, the root tip was actively growing. It was cut off, sectioned, stained, and preserved with a resin under the cover slip. Consequently it is a great area to observe onion cells in various phases of mitosis and cell division.
2. Set up your microscope and obtain an *Allium* root tip slide. Place it on the stage and secure it with stage clips. Look at the root tip (longitudinal section) with the naked eye to remind yourself how small it actually is. Position it so that the root tip is directly over the hole in the stage. Start with the scan power objective and focus.
3. Switch to low power (focus it) and then switch to high power (focus it). Find cells in interphase and the four phases of mitosis: prophase, metaphase, anaphase, and telophase. Also, don't view the root cap. View the area of actively dividing cells covered by the root cap. (The root cap is a protective thimble-shaped group of cells covering the root tip.) To make sure you can properly identify each phase of mitosis by their appearance, visit http://logospressonline.com/content/RiotLinks.pdf, click on the *second* Laboratory 10 link, and look at the labeled images.

D. Acting out mitosis and cell division

1. Each embroidery floss skein (of the twelve) should have double-sided tape wrapped around its middle. Stick together the skeins that are the

same color. You should have six **duplicated chromosomes** (two-copy chromosomes): Each chromosome consists of two skeins of the same color: two light blue stuck together, two dark blue, two red, two pink, two dark green, and two light green.

2. Put these six two-copy embroidery-floss chromosomes in one of the transparent trash bags. This represents the cell's nucleus, and the room represents the cell. Pretend that all the embroidery floss is not nicely bundled up into skeins. Imagine the floss is totally uncoiled throughout the entire nucleus. This would be Interphase.

Prophase

3. Hold the bag in the middle of the room. Imagine that the uncoiled floss (DNA) is coiling up into nice tidy skeins (chromosomes).
4. Next, dump the six stuck-together "two-copy chromosomes" out of the bag on to the floor and stash the bag out of sight. This represents the nuclear envelope disintegrating and releasing the chromosomes into the cytoplasm.
5. Get a partner. You and your partner should face each other, each of you holding a (two-copy) chromosome (you two represent the mitotic spindle, since you are moving the chromosomes around). Move the chromosomes around, back and forth, and eventually bring all these chromosomes toward the **plane of division** which should be in the middle of the room at a right angle to the long axis of the room. It is still prophase until the chromosomes are lined up.

Metaphase

6. Once the two-copy chromosomes are lined up on the plane of division, the cell is in metaphase. Have the each chromosome oriented so that their copies are facing towards opposite sides of the cell.

7. How many chromosomes does this cell have? ______________ Are the six chromosomes one or two-copy? _______________

Anaphase

8. You and your partner should pull apart each two-copy chromosome. Each of you is now holding one-copy of each chromosome. **You are not holding half a chromosome.** This begins anaphase. You and your partner should slowly walk away from each other to opposite ends of the room, carrying your one-copy chromosomes. You should move to one end of the room and your partner should move to the other end of the room. Each of you place your one-copy chromosomes in a single pile at opposite ends of the room.

Telophase

Once the chromosomes stop moving telophase has begun.

9. How many chromosomes are in the pile? ___________________________
10. Are they one or two-copy?____________________________________
11. Did this division reduce the number of chromosomes? ________________
12. Each of you should take a plastic trash bag and insert your six chromosomes. This represents the nuclear envelope reforming around the chromosomes. The spindle microtubules can disintegrate because their purpose of hauling the chromosomes around is over.
13. Pretend the chromosomes within the bag uncoil to complete telophase.
14. Are these two new cells ready to do mitosis right now?________________ Why? ____
15. The starting mother cell had six two-copy chromosomes. Now there are two daughter cells, each having _______________ one-copy chromosomes.
16. What needs to occur for each daughter cell to be ready to do mitosis again?

____

17. Why is S phase of interphase important before the cell begins mitosis?

____

Cytokinesis

Cytokinesis happens during telophase. The mother cell separates into two daughter cells.

18. Place a row of chairs (or some other prop) to split the room into two halves.
19. Discuss the cleavage furrowing in animal cell cytokinesis and cell plate formation in plant cell cytokinesis.
20. What are the differences? ______________________

____

LABORATORY 11

MEIOSIS

MATERIALS

- A computer to watch an online video.
- Compound microscope
- Lily anther meiosis slide
- Same twelve thin skeins of embroidery floss used in the last lab
- Four clear trash bags

PREPARATION

Make sure you have read Chapter 13 "Meiosis" in *The Riot and the Dance* (pp. 153–160).

OBJECTIVES

i. Know the definition of meiosis, homologous chromosomes, crossing over, diploid, haploid, reduction division, gamete, and spore.

ii. Know the reason for meiosis.

iii. Know where it occurs in plants and animals and what it produces in each.

iv. Be able to identify each phase of meiosis in the lily anther.

v. Be able to describe the process of meiosis and the phases within it:

- Meiosis I: prophase I, metaphase I, anaphase I, telophase I
- Meiosis II: prophase II, metaphase II, anaphase II, telophase II

EXERCISES

A. Definitions

meiosis: ____________________

homologous: ____________________

crossing over: ____________________

diploid: ____________________

haploid: ____________________

reduction division: ____________________

gamete: ____________________

spore: ____________________

B. B. Video demonstration

Visit http://logospressonline.com/content/RiotLinks.pdf. Click the link for the Laboratory 11 video, "Meiosis," and watch the whole thing (to 19:25). This is what you'll be acting out in Section D.

C. Examination of cells undergoing meiosis

The lily anther is where pollen grains form. Each grain came from a haploid microspore that was formed by meiosis in the anther. When the slide was made, cells were in the process of meiosis to make haploid microspores. In the process it was sectioned, stained, and preserved with a resin under the cover slip.

1. Set up your microscope and obtain a slide of a lily anther labeled meiosis. Place it on the stage and secure it with stage clips. Look at the anther (cross section) with the naked eye to remind yourself how small it actually is. Position it so that the anther is directly over the hole in the stage. Start with the scan power objective and focus.
2. Switch to low power (focus it) and then switch to high power (focus it). Find cells in the eight phases of meiosis I and II. To make sure you can properly identify each phase of meiosis by its appearance, visit http://logospressonline.com/content/RiotLinks.pdf, click on the second link for Laboratory 11, scroll down to "meiosis under a microscope..," and look at the black and white, labeled microscopic images.

D. Acting out meiosis

1. Stick the same-colored skeins together in pairs with the double-sided tape as you did in the last lab. When they are stuck together, they are **two-copy chromosomes** (or duplicated chromosomes). Put them all in one of the transparent trash bags. This represents the cell's nucleus, and the room represents the cell. Pretend that all the embroidery floss is not nicely bundled up into skeins. Imagine the floss is totally uncoiled throughout the entire nucleus. This would be Interphase.

Prophase I

2. Hold the bag in the middle of the room. Imagine that the uncoiled embroidery floss (DNA) is coiling up into nice tidy skeins (chromosomes).
3. Next, dump the six skeins out of the bag on to the floor and stash the bag for later. This represents the nuclear envelope disintegrating and releasing the chromosomes into the cytoplasm.
4. Get a partner. Now, **homologous chromosomes** are represented by the pair of two-copy chromosomes that have different shades of the same color. One

of you should be holding one of each pair of homologous chromosomes, and the other holding the matching two-copy chromosomes. In other words, one of you is holding the light blue/pink/light green chromosomes and the other is holding the dark blue/red/dark green next to them—which makes three pairs of homologous chromosomes. Imagine the homologous pairs swapping equivalent pieces of their chromosomes (too difficult to actually do this with the embroidery floss). This is called **crossing over.** Again you and your partner represent the mitotic spindle jockeying the chromosomes around. Walk around (but still keep the homologous pairs together), back and forth, and eventually bring the three pairs of homologous chromosomes toward the **plane of division,** which should be in the middle of the room, at a right angle to the long axis of the room. It is still prophase I until the chromosomes are lined up.

Metaphase I

5. Once the homologous pairs are lined up on the plane of division, the cell is in metaphase I. Have each chromosome oriented so that one from each homologous pair is facing towards opposite sides of the cell.
   How many chromosomes does this cell have?________________________
   Are the six chromosomes one- or two-copy?________________________
   How many homologous pairs are there? ________________________

Anaphase I

6. You and your partner should now separate the homologous pairs (not the copies that are stuck together). This begins anaphase I. As you and your partner move to opposite sides of the room you are still holding two copy chromosomes. Each of you should place your two-copy chromosomes in a single pile on opposite ends of the room. If you did this correctly, each side of the room should have three two copy chromosomes (one from each

homologous pair; one from the red pair, one from the green pair, one from the blue pair).

Telophase I

7. Once the chromosomes stop moving telophase I has begun.
   How many chromosomes are in each pile? ____________
   Are they one- or two-copy? ____________
   We started with a diploid cell; what is the ploidy of each cell now? _______
8. Since this division reduced the number of chromosomes it is called the ________________________ division.
9. Take two plastic trash bags and have you and your partner put each group of three chromosomes in each bag. This represents the nuclear envelope reforming around the chromosomes during Telophase I. The spindle microtubules can disintegrate because their purpose of hauling the chromosomes around is over.
10. Pretend the chromosomes within the bag uncoil to complete telophase I.
11. Why are these two new cells haploid and not diploid?________________
    ____
12. Since there is one blue, one red (or pink), and one green in each cell, we know that each is haploid (even though each chromosome is two copy). Diploid cells would have two blues, two reds, and two greens.
13. The starting mother cell had six two-copy chromosomes. Now there are two daughter cells, each having ____ two-copy chromosomes. This is why the ploidy (haploid or diploid) of the cell has nothing to do with how many copies each chromosome has. After mitosis, both daughter cells have one copy chromosomes and they are still diploid. After meiosis I the daughter cells have two-copy chromosomes, and they are haploid. Ponder this until you're not confused (or maybe watch the video again).

Cytokinesis

Cytokinesis happens during telophase I. The mother cell separates into two daughter cells.

14. Place a row of chairs (or some other prop) to split the room into two halves.

Meiosis I is finished.

Meiosis II begins.

Start Prophase II

15. In the two separate daughter cells, dump the three skeins out of each bag on to the floor and stash the bag out of sight. This again represents the nuclear envelope disintegrating and releasing the chromosomes into the cytoplasm.
16. This division will be more like mitosis because we will be separating the two copies. You and your partner are each in a different daughter cell this time (separated by a row of chairs), acting out Meiosis II.
17. Each of you hold the (two-copy) chromosomes in your respective cell (again, you and your partner represent the mitotic spindle in each daughter cell, since you are moving the chromosomes around). Move the chromosomes around, back and forth, and eventually bring all these chromosomes toward the **plane of division** (which should be in the middle of each daughter cell at right angles to the first plane of division of Meiosis I, also known as the row of chairs). It is still prophase II until the chromosomes are lined up.

Metaphase II

18. Once the three (two-copy) chromosomes are lined up on the plane of division in each cell, the cell is in metaphase II. Have each chromosome oriented so that their copies are facing towards opposite sides of the cell.

How many chromosomes does this cell have? ____________

What is the ploidy? haploid or diploid (circle one) ?

Anaphase II

19. In your respective cells, each of you should pull apart your three two-copy chromosomes. Now you are now holding three one-copy chromosomes in each hand (since you're each doing this alone you might stuff the three one-copy chromosomes between your fingers and do the same for the other hand). **You are not holding three half chromosomes in each hand.** This begins anaphase II. As before, carry the three one-copy chromosomes to opposite sides of the cell (you'll have to drop off the chromosomes in your left hand and then walk back to drop off the chromosomes in your right, which of course isn't how cells do it, but it's close enough). Your partner should be doing the same thing on the other side of the room. Place the one-copy chromosomes in a single pile at both ends of each cell.

Telophase II

20. Once the chromosomes stop moving telophase II has begun.

    How many chromosomes are in each pile? __________

    Are they one or two-copy? __________

    Since there is one blue, one red (or pink), and one green in each cell, we know that it is haploid.

    Since meiosis II had cells go from **three** (two-copy) chromosomes to **three** (one-copy) chromosomes, did this division reduce the number of chromosomes? __________

21. Take four plastic trash bags and put the 4 groups of three chromosomes in each. This represents the nuclear envelope reforming around the

chromosomes. The spindle microtubules can disintegrate because their purpose of hauling the chromosomes around is over.

22. Pretend the chromosomes within the bag uncoil to complete telophase II.

Cytokinesis

Cytokinesis happens during telophase II. Each cell from meiosis I separates into two daughter cells. The plane of division should be at right angles from the first division.

23. Place a row of chairs (or some other prop) to split each daughter cell of meiosis I in half. Now the room should be divided into four cells.

This is the process to produce gametes (sperm and eggs) in humans and animals. It is also the process to produce spores in plants.

LABORATORY 12

MENDELIAN GETICS

MATERIALS

- Paper and pen to work various crosses using Punnett squares

PREPARATION

Make sure you have read Chapter 14 "The Basics of Mendelian Genetics," in *The Riot and the Dance* (pp. 163–172).

OBJECTIVES

i. Know the definitions of allele, dominant, recessive, homozygous, heterozygous, Punnett square, diploid, haploid, genotype, phenotype, Law of segregation, and Law of independent assortment

ii. If given the genotype of parents, be able to determine types of gametes possible. Also be able to cross them using a Punnett square to figure out the genotypes and phenotypes of the offspring and the proportions of each.

EXERCISES

A. Definitions

Allele: ___

Dominant and recessive: ___

Homozygous: ___

Heterozygous: ___

Punnett square: ___

Diploid (review): ___

Haploid (review): ___

Genotype: ___

Phenotype: ___

Law of Segregation: ___

Law of Independent Assortment: ___

B. Practice crosses

Mendel's Pea Plants

Use the set of pea traits below to answer the following questions.

Y = *yellow seeds (dominant)*

y = *green seeds (recessive)*

R = *round seeds (dominant)*

r = *wrinkled seeds (recessive)*

1. F1: Yy x Yy (monohybrid cross)

   What is the proportion of green seeds in the F2 generation?

   ___

2. F1: YyRr x YyRr (dihybrid cross)

   What is the proportion of F2 offspring have green round seeds?

   ____

   What is the proportion of F2 offspring have yellow wrinkled seeds?

   ____

Hydras

Use the set of hydra traits below to answer the following questions.

F = *fanged*

f = *fangless*

C = *ten-headed**

c = *two-headed**

*C and c exhibit incomplete dominance. Heterozygotes have six heads.

3. F1: FfCc x FfCc What is the phenotype of these parents?

   What proportion of F2 offspring don't have fangs and have six heads?

   ____

   What proportion of F2 offspring don't have fangs and have ten heads?

   ____

   What proportion of F2 offspring have fangs and have two heads?

   ____

LABORATORY 13

CLASSIFICATION

MATERIALS

- A computer to watch an online video.
- Pressed and mounted leaves (unless fresh are available) of the following kinds: maple, lupine, black locust, mountain ash, beech, cattail, and magnolia (if actual leaves are unavailable use online images). Have each plant's identity available (but not visible to the student) so they can know if they keyed it correctly.
- Preserved centipede, millipede, beetle, butterfly, dragonfly, grasshopper, and roly-poly.

PREPARATION

Make sure you have read Chapter 15 "Classifying Life" in *The Riot and the Dance* (pp. 179–200).

OBJECTIVES

i. Know who Carolus Linnaeus is and the seven main levels of classification in the Linnaean hierarchy.

ii. Know how to correctly write a scientific name.

iii. Know how to use a dichotomous key.

iv. Know how to create a simple dichotomous key

EXERCISES

A. Definitions:

Use the textbook glossary to define the following terms, and think about the definitions while doing this exercise.

Carolus Linnaeus: ______________________

kingdom: ______________________

phylum:______________________

class: ______________________

order: ______________________

family:______________________

genus: ______________________

species: ______________________

scientific name (synonyms—species name, Latin name, and binomial):

____

Specific epithet: ______________________

B. Using a key.

Use the key to figure out the identity of the seven pressed plants. The six pairs of statements to which each specimen is compared are called *couplets*. Below the key are some descriptive terms to help you key these plants.

Useful Terms for Keying Selected Plants

Compound leaf—the leaf blade is divided up into leaflets which look like small leaves

- *Palmately compound*—leaflets radiate from a single point
- *Pinnately compound*—leaflets branch off of either side of the central mid-vein (rachis) with a terminal leaflet at the end.

Simple leaf—the leaf is not divided up into leaflets and looks like a single leaf

Venation—the pattern of veins in a leaf

- *Parallel venation*—all main veins run parallel to each other.
- *Pinnate venation*—the pattern of veins in the leaf have a central mid-vein with lateral veins branching off of either side
- *Palmate venation*—the pattern of veins are more like spokes in a wheel; several or many veins radiate out from one central point near the base of the leaf

Serrate—sawlike edge to the leaf or leaflet.

Entire—smooth leaf edge

Key for Identifying Selected Plant Species

1. a. Leaves simple ....go to 2
   b. Leaves compound ....go to 5
2. a. Leaf venation parallel..Cattail
   b. Leaf venation not parallel ....go to 3
3. a. Leaf with palmate venation..Maple
   b. Leaf with pinnate venation..go to 4
4. a. Leaf serrate .... Beech
   b. Leaf entire..Magnolia
5. a. Palmately compound..Lupine
   b. Pinnately compound ....go to 6
6. a. Leaflets serrate ....Mountain Ash
   b. Leaflets entire .... Locust

C. Making a Key

Make your own key for the following arthropods: centipede, beetle, butterfly, millipede, dragonfly, and grasshopper. Below is a short list of characteristics that will help you make a few couplets to allow someone to successfully key these few arthropods. There can only be two choices in a couplet (that's why it's called a couplet). Take some time to closely observe these creatures to come up with good couplets. Although not necessary, the goal is to make an effective key with as few couplets as possible. Refer to the plant key above for ideas on how to phrase your couplets.

- Number of legs.

  A couplet for this could be:

  a. Animal with six legs .... go to X.

  b. Animal with greater than six legs ....go to Y.
- Length of body
- Size of legs
- Type of wings or wing covers

D. Writing a scientific name.

7. What is the common name of your favorite animal and plant? Recall that a scientific name is synonymous with Latin name and species name. Look them in an encyclopedia/dictionary or online (e.g., their Wikipedia entry) and find out their scientific names.

   For example:

   Common name: Common box turtle
   Scientific name: *Terrapene carolina*

   Common name: Coast redwood
   Scientific name: *Sequoia sempervirens*

Common name: ______________________________

Scientific name: ______________________________

Common name: ______________________________

Scientific name: ______________________________

What is the genus of your favorite plant? ______________________________

What is the specific epithet of your favorite plant? ______________________________

What is the species name of your favorite plant (careful, I'm not asking the same question twice)? ______________________________

LABORATORY 14

KINGDOM PROTISTA

MATERIALS

Suggested vendor: Carolina Biological Supply Company

- Compound microscope
- Glass slides and cover slips
- Spirogyra (living)
- Protozoa mixture (living)
- Protoslo®

PREPARATION

Make sure you have read Chapters 17 and 18, "The Algae: Plant-like Protists" and the "The Fungal-like and Animal-like Protists," in *The Riot and the Dance* (pp. 209–218; 221–232). Just a quick "middle of the year" encouragement: Try not to be in "checklist mode" or "do what the lab book says" mode. Yes, I want you to look at these things, but it's better to indulge and jump start your sense of wonder and curiosity looking at fewer specimens than to get through all the lab exercises (because you're dutiful and diligent) bereft of a love of learning. Don't let your sense of wonder wither. These are all the handiwork of God. Ponder their tiny size and amazing complexity. If you are mesmerized and like what you see, then thank the Creator while sitting at the microscope.

OBJECTIVES

i. Know the characteristics (or lack of characteristics) that exclude protists from the other kingdoms.
ii. Know what phylum each specimen belongs in and why they have been placed in that phylum by taxonomists.
iii. Become more proficient at using a microscope.
iv. Enjoy watching *Paramecium*, *Stentor*, and *Spirogyra* and also know their basic structure.
v. Hone observational and drawing skills in examining these protists under the microscope.

EXERCISES

A. Definitions:

Use the textbook glossary to define the following terms, and think about the definitions while doing this exercise.

pseudopod:______________________________

oral groove: ______________________________

food vacuole: ______________________________

contractile vacuole: ______________________________

cilia: ______________________________

pellicle: ______________________________

chloroplast: ______________________________

B. Paramecium (member of phylum Ciliophora)

1. Get a blank slide. Using a plastic dropper, suck up a piece of "skuzzy material" (about a drop) from the bottom of the *Paramecium* jar and place it in the middle of the slide. The drop should have plenty of paramecia in it.

2. Sparingly apply a circular bead of Protoslo around the drop. This syrupy substance will slow down the speedy paramecia so that you can view them more easily.
3. Place the edge of a cover-slip on the edge of the drop and slowly lower it over the drop so that the water spreads out evenly under the cover-slip and doesn't trap air bubbles. The Protoslo will diffuse inward and impede the paramecia that get engulfed by it. (Recall from the beginning of the year that steps 1-3 are referred to as a **wet mount.**)
4. Place the slide on the stage and secure it with stage clips. Center the patch of "skuzzy material" so that it's positioned directly above the hole in the stage where the light shines through.
5. View at scan power (always start there). Focus as you did before (always coarse, then fine), bringing the patch of skuzz into focus. Look for torpedo-shaped paramecia. There should many to pick from.
6. Once you found one that is moving slowly or sitting still, center it in the middle of the *field of view* and rotate the nosepiece to low power. Watch it move about using its hair-like **cilia.** Adjust the lighting using the iris diaphragm lever so that it's not too bright or not too dark.
7. Using your fingers to move the slide, center the *Paramecium* and rotate the nosepiece to high power. (**Careful: at high power the objective will be very close to the slide so don't use the coarse adjustment because you could easily bash the lens into the slide.**) Using **fine adjustment** bring it into sharp focus. Again, adjust lighting using the iris diaphragm lever. Now watch it move for as long as you like. Enjoy. These cells look like living transparent kayaks. They are completely covered in **cilia.** If you are on high power and have it in good focus and lighting, you should be able to see the beating of thousands of cilia. There is a groove running along the side that you should notice as the paramecium spins. This is called

the **oral groove**. Bacteria and other smaller unicellular critters are swept down the groove into a food vacuole. Note the cell maintains a certain shape. A **pellicle** is a cytoskeletal structure under the cell membrane of the *Paramecium*. If you are lucky, you'll see a clear spherical area that repeatedly shrinks and enlarges. That is the **contractile vacuole.** Since the paramecium is a freshwater inhabitant, it is surrounded in a hypotonic environment. This means that water will be constantly flowing into the cell by osmosis. Since paramecia don't have cell walls they are in danger of bursting (cytolysis). Contractile vacuoles collect the excess water flooding into the cell and expel it through a canal running between the vacuole and the cell membrane.

8. Take a few minutes to sketch the *Paramecium* (hopefully it's somewhat stationary). Pay attention to detail. Try to capture the overall shape and contours of the oral groove. Label the structures mentioned above. As it moves you might have to move the slide with your fingers to keep it in view.

9. Refer to your calculation of the diameter at high power (D2) in Laboratory 1. What is the approximate size of the *Paramecium* you drew?

   __________ mm __________ µm

C. Stentor (member of phylum Ciliophora)

1. In the same way as *Paramecium*, do a wet mount of *Stentor* and observe it closely. Again start at scan power and find an individual *Stentor* that is not swimming. Focus and adjust lighting. Rotate to low power, adjust everything again and rotate it to high power. Enjoy.

   Is it bigger than *Paramecium*? ______________

   Can it change shape? ______________

   Describe its behavior. ____

   ____

   ____

   ____

2. Sketch *Stentor* and label the pellicle (this will look like longitudinal pin-striping), oral groove, and cilia. Note that the cilia are longer on the rim of the funnel-shaped opening.

D. Spirogyra (member of phylum Chlorophyta)

1. Get a blank slide. Using a plastic dropper, suck up a few green filaments of the algae called *Spirogyra* along with a drop of water and place it in the middle of the slide.

2. Place the edge of a cover-slip on the edge of the drop and slowly lower it over the algae-containing drop so that the water spreads out evenly under the cover-slip and doesn't trap air bubbles. If the algal filaments are too long, trim any that are sticking out so they fit on the slide and under the cover slip

3. Place the slide on the stage and secure it with stage clips. Center the algal filaments so that they are positioned directly above the hole in the stage where the light shines through.

4. View at scan power. Focus as you did before (always coarse, then fine), bringing the green filaments into focus.

5. Center some healthy-looking green filaments in the middle of the field of view and rotate the nosepiece to low power. Adjust the lighting using the iris diaphragm lever so that it's not too bright or not too dark.

6. Rotate the nosepiece to high power. Center a few rectangular-shaped cells (really cylindrically-shaped) and take a few minutes to sketch at least two adjacent cells. Pay close attention to detail. Use a bracket to label one cell. Note and label the outermost **cell wall**, the **spiral-shaped chloroplasts** running the length of the cell, and the centrally located **nucleus**.

LABORATORY 15

KINGDOM FUNGI

MATERIALS

Suggested vendor: Carolina Biological Supply Company

- Compound microscope
- *Rhizopus* with sporangia and zygosporangia (prepared slide)
- Moldy bread (prepare a week or more in advance)
- Cup fungus (*Peziza*) (prepared slide)
- Dried ascomas—cup fungus and morel
- *Penicillium* (prepared slide)
- Mushroom (*Coprinus*) (prepared slide)
- Variety of dried basidiomas (purchased or personally collected)—mushroom, bolete, shelf fungus, puffball, earthstar, coral fungus (Don't worry if you don't have them all. It's just good to get a sense of the diversity of forms out there.)
- Lichen (*Physcia)* (prepared slide)
- Crustose, foliose, and fruticose lichens

PREPARATION

Make sure you have read Chapter 19 "Kingdom Fungi" in *The Riot and the Dance* (pp. 235–251). Keep your sense of wonder fired up as you survey the major groups of fungi. These are amazing life forms that not only feed on dead (or occasionally living) organic matter, but also add a bizarre beauty to many shady and moist forests. Enjoy!

OBJECTIVES

i. Know the characteristics of Kingdom Fungi
ii. Become more proficient at using a microscope.
iii. Know the basic characteristics of the phylum Zygomycota using the life cycle of the black bread mold (*Rhizopus*).
iv. Know the basic characteristics of the phylum Ascomycota using the life cycle of the cup fungus (*Peziza*). Know the asexual structures of *Penicillium* mold. Know why morels belong to this phylum.
v. Know the basic characteristics of the phylum Basidiomycota using the life cycle of a typical mushroom (*Coprinus*). Know the various representatives on demonstration that show the diversity within this phylum.
vi. Know the basic structure and the three growth forms of lichen.
vii. Hone observational skills in examining these fungi under the microscope.

EXERCISES

A. Definitions:

Use the textbook glossary to define the following terms, and think about the definitions while doing this exercise.

stolon: ______________________________

sporangiophore : ______________________________

sporangium: ______________________________

spores: ______________________________

rhizoids: ______________________________

zygosporangium: ______________________________

ascoma: ______________________________

ascus: ______________________________

ascospores: ______________________________

conidiophore: ______________________________

conidia: ______________________________

basidioma: ______________________________

basidium: ______________________________

basidiospores: ______________________________

hymenium: ______________________________

lichen: ______________________________

B. Phylum Zygomycota (*Rhizopus*)

1. Get a prepared slide of *Rhizopus* (Black Bread Mold) and examine it under the microscope (scan, low, and high power). Note any hyphae that end in a dark colored knob. The knob is the **sporangium** which contains **spores.** The hypha that leads up to the sporangium is the **sporangiophore** (sporangium-bearer). Using high power sketch all three below and label them.

2. Look around using scan power and find some branching root-like hyphae. These are **rhizoids.** Any hyphae on either side of the base of the rhizoids that don't lead to sporangia are **stolons.** Try to find these hyphae. Also find a **Zygosporangium** which looks like a dark bumpy sphere held between two hyphae that were growing toward each other. Sketch and label it below.

C. **Phylum Ascomycota** (*Peziza*)

1. Mount a prepared slide of a cup fungus (*Peziza*) and examine it under the microscope (scan → low power). Note the layer lining the inner surface (**hymenium**) of the cup-like fruiting body (**ascoma**). Note all the parallel **asci** forming the hymenium. The asci are at right angles to the inner surface of the cup.

3. Center the asci in the field of view, switch to high power, and then focus. Examine one well defined **ascus** containing eight **ascospores.** Sketch one ascus containing the eight ascospores.

4. Examine a prepared slide of *Penicillium.* There is also a slice of whatever it was growing on. Look at the edge of the object using scan power. Center it and focus. Switch to low power. Center it and focus. Then switch to high

power, center it and focus. Note the several hyphae branching from one point with tiny spores arranged like pop beads on the tips of the branches. The spores are **conidia** and the branching hyphae bearing them are **conidiophores**. Sketch and label both.

D. **Phylum Basidiomycota** (*Coprinus*)

1. Mount a prepared slide mushroom (*Coprinus;* cross section of its cap) and examine it under the microscope (scan → low power). Note the gills will appear like spokes in a wheel. The edge of the cap looks like the wheel and the stipe looks like the axel. Look at the layer lining the gills (**hymenium**) of the fruiting body (**basidioma**). Note all the parallel **basidia** forming the hymenium. The basidia are at right angles to the surface of the gills. ("Basidioma" is the technical name of the entire mushroom.)

5. Center the basidia in the field of view, switch to high power, and then focus. Examine one well defined **basidium** with four **basidiospores** perched on the outer surface of the basidium. You might not see all four since the back two might be hiding behind the front two. Sketch one basidium with its basidiospores and label them.

6. Look at the two ascomas (cup fungus and morel) and several basidiomas on display. Be able to identify the type of fungus and what phylum it belongs in. For example, "That's a puffball and it belongs to Phylum Basidiomycota."

E. Lichens

1. Mount a prepared slide of lichen (*Physcia*) and examine it under scan, low, and high power. Remember to center a good area of the specimen at scan power and focus it before switching to the next power. Note the bluish green fungal hyphae (Ascomycota) forming a matrix in which the *reddish algal cells* (Chlorophyta) are embedded. The algal cells are not their natural color due to the staining technique. The upper layer of the lichen is a dense mesh of hyphae called the *upper cortex*. The middle of the lichen is a loose mesh of hyphae called the *medulla* and the lower layer is also dense and is called the *lower cortex*. The reddish oval algal cells are mostly in the upper cortex and the medulla near the upper cortex. Sketch a small section of the lichen labeling these parts.

2. Examine the three main growth forms of lichen. In your own words how would you describe the shape of each?

   Crustose: ____

   Foliose: ____

   Fruticose: ____

LABORATORY 16

PHYLA PORIFERA AND CNIDARIA

MATERIALS

Suggested vendor: Carolina Biological Supply Company

- Compound microscope
- Depression slides
- Cover slips
- *Scypha* longitudinal section (prepared slide)
- Bath sponge (dead skeleton)
- Spongin (prepared slide)
- Spicules (prepared slide)
- *Hydra* longitudinal section (prepared slide)
- Live *Hydra*
- Live *Daphnia*
- Preserved jellyfish
- Preserved sea anemone
- Coral skeleton

PREPARATION

Make sure you have read Chapter 20 "Kingdom Animalia: A Short Introduction," Chapter 21 "Phylum Porifera," and Chapter 22 "Phylum Cnidaria" in *The Riot and the Dance* (pp. 255–256; 259–262; 265–272). We are starting our survey of the major groups of animals. Most specimens will be dead, preserved specimens or prepared slides, but some will be alive so our labs won't be entirely without movement. Of course, the more you see these things in the great outdoors, the more this lab will seem like looking at a catalog rather than actually going shopping. The point of this is to inspire you to be more curious and observant when you do go on an outing, whether on land or sea.

OBJECTIVES

i. Know the characteristics of Kingdom Animalia.
ii. Become even better at using a microscope.
iii. Know the basic characteristics of phylum Porifera.
iv. Know the basic structure of a simple sponge (*Scypha*) and the path of water flow.
v. Know the basic characteristics of phylum Cnidaria and its two body forms: the polyp and medusa.
vi. Know the basic structure of *Hydra* and how it feeds on *Daphnia*.

EXERCISES

A. Definitions

Use the textbook glossary to define the following terms.

osculum: ______________________________

spongocoel: ______________________________

incurrent canal: ______________________________

radial canal: ______________________________

prosopyles: ______________________________

collar cells: ______________________________

polyp: ____________________

medusa: ____________________

epidermis: ____________________

cnidocytes: ____________________

nematocysts: ____________________

gastrodermis: ____________________

mouth: ____________________

gastrovascular cavity: ____________________

pedal disc: ____________________

jellyfish: ____________________

sea anemones: ____________________

coral: ____________________

B. *Scypha* **(a synconoid sponge)**

1. Mount a prepared slide of *Scypha* (longitudinal section) and examine it under the microscope (scan, low, and high power). Sketch the upper part of the body (on scan power) and label an incurrent canal, prosopyles, radial canals, spongocoel, and osculum. Use arrows to draw the flow of water through the sponge.

2. Examine other sponge specimens on demonstration and note the differences in size and shape.

C. *Hydra* (polyp)

1. Mount a prepared slide of *Hydra* (longitudinal section) and examine it under the microscope (scan, low, and high power). Sketch the upper third of the body and label the epidermis, gastrodermis, gastrovascular cavity, mouth, and tentacles.

2. Observe *Hydra* feeding. Using a dropper prepare a wet mount of one or two living *Hydra* (along with a drop of water) in the slight concavity of a depression slide. Place a few small *Daphnia* crustaceans in with the *Hydra* (don't get too much water on the slide). Place the cover slip over the drop and view at scan power on the microscope. It won't be long before one or more of the *Daphnia* blunder into the stinging tentacles of the *Hydra*. The nematocysts will fire (they are too small to see but you'll see the effects), stinging and capturing the tiny crustacean. Over the course of a half hour or so the *Hydra* will open wide its mouth and swallow the entire *Daphnia* whole. In a short paragraph below describe the event.

D. Diversity of Cnidaria

1. Observe a preserved jellyfish in a dish of water. Note the tentacles and mouth on the underside.
   What body form does the jellyfish have? ______________________________

2. Observe a preserved sea anemone in a dish of water. Note the tentacles and mouth on the upper side. What body form does the sea anemone have?

   ____

3. Observe a coral skeleton. Note the many small cavities pitting the entire surface of the coral rock. These are the holes which were occupied by coral animals when the coral was alive. Coral animals look like miniaturized sea anemones in a stone apartment complex. What body form do the coral animals have? ______________________________

LABORATORY 17

THE WORMS

MATERIALS

Suggested vendor: Carolina Biological Supply Company

- Compound microscope
- Live *Planaria*
- Empty Petri dish (or other shallow dish)
- Sheep liver fluke whole mount (prepared slide)
- *Dipylidium canium* scolex and proglottids (prepared slide)
- Live vinegar eels (*Turbatrix aceti*)
- Live earthworm
- Preserved leech
- Preserved *Nereis* (clamworm)

PREPARATION

Make sure you have read Chapter 23 "The Worms" in *The Riot and the Dance* (pp. 275–280). As we continue our survey the major groups of animals we will have a brief tour of three phyla of worms: Phylum Platyhelminthes, Phylum Nematoda, and Phylum Annelida. Worms may not seem very interesting, but as you study and observe them you'll be surprised at the amazing variety of morphology and natural history.

OBJECTIVES

i. Know the characteristics of Phylum Platyhelminthes (the flatworms) and examples of the three major groups (one free-living and two parasitic). Be able to recognize a typical member of each group, e.g., a planarian, a fluke, and a tapeworm.

ii. Know the characteristics of Phylum Nematoda (the roundworms) and two examples of parasitic life cycles. Be able to recognize a round worm.

iii. Know the characteristics of Phylum Annelida (the segmented worms) and the three major groups. Be able to recognize a typical member of each group, e.g. earthworm, leech, and clamworm.

EXERCISES

A. Definitions

Use the textbook glossary to define the following terms.

Flukes: ____________________

Tapeworms: ____________________

Scolex: ____________________

Proglottid: ____________________

Planarian: ____________________

Nematodes: ____________________

Segmented worms: ____________________

Clamworm: ____________________

B. Phylum Platyhelminthes (the flatworms)

1. What are the main characteristics of Phylum Platyhelminthes?

____

____

____

2. Place one *Planaria* in a shallow dish of water with a few crumbs of a crushed dog food nugget. Observe and describe the feeding and movement of *Planaria*. Sketch it and label the head, eyespots, and pharynx.

3. Mount a prepared slide of a sheep liver fluke and a tapeworm and observe them closely. Name two structures flukes have that tapeworms don't.

____

____

4. Name two structures tapeworms have that flukes don't.

____

____

What flatworm class do flukes belong in? ____________________

What flatworm class do tapeworms belong in? ____________________

C. Phylum Nematoda (the roundworms)

Although there is much diversity in size and natural history most have a remarkably similar shape. Observe and briefly describe the behavior and overall shape of the live vinegar eels (small, free-living nematodes).

____

____

____

How do they differ from flatworms in their shape? ____________________

____

D. Phylum Annelida (the segmented worms)

1. Earthworm: Observe a large live earthworm for a few minutes in a container with a shallow bedding of moist soil. Describe its thickness or girth where it lengthens and where it shortens.

   Which muscles do you think are contracting when the earthworm lengthens: circular or longitudinal muscles? ____________________

   Which muscles do you think are contracting when it shortens: circular or longitudinal muscles? ____________________

   What do earthworms consume? ____________________

2. Hold the worm between finger and thumb. Using your other finger and thumb, gently slide them down the length of the earthworm.

   What do you feel? ______________________________

   What's their function? ______________________________

3. Leech and Clamworm: Observe these two preserved annelids.

   What external feature do they both share with earthworms? In other words, why are they considered annelids? ______________________________

   Leeches have two suckers. Where are they located? ______________________________
   Look closely at the fleshy appendages of the clamworm. Do you see tufts of bristles at the tips of the appendages?______________________________

   Are they longer and more numerous than the bristles in the earthworm?

   ____

LABORATORY 18

PHYLUM MOLLUSCA

MATERIALS

Suggested vendor: Carolina Biological Supply Company

- Basic dissection kit (scalpel, dissection needle, dissection probe, scissors, and tweezers)
- Dissection tray / latex gloves
- Empty petri dish (or other shallow dish)
- Preserved clam (4 inches long)
- Live leopard slug
- Preserved squid or octopus
- An assortment of different bivalve shells, gastropod shells, and a chambered nautilus shell.

PREPARATION

In preparation for Lab 18 read carefully Chapter 24 "Phylum Mollusca" in *The Riot and the Dance* (pp. 283–289). Mollusks are a very diverse phylum ranging from sedentary clams to speedy squid. As we continue our survey of the major groups of animals we will get acquainted with some of the main representatives of the major classes of mollusks. Again you'll be amazed at the huge disparity in morphology and natural history.

OBJECTIVES

i. Know the six major characteristics of Phylum Mollusca and examples of the three major classes: Classes Bivalvia, Gastropoda, and Cephalopoda. Be able to name the class a particular mollusk belongs in just by sight.
ii. Know the basic anatomy and general feeding habits of representative mollusks that are described in the textbook.

EXERCISES

A. Definitions

Use the textbook glossary to define the following terms.

mantle: ____________________

mantle cavity: ____________________

visceral mass: ____________________

head-foot: ____________________

incurrent siphon: ____________________

excurrent siphon: ____________________

radula: ____________________

pneumostome: ____________________

ink sac: ____________________

B. Class Bivalvia

1. Bivalves are filter-feeding mollusks with two shells. Examples are clams, oysters, cockles, mussels, scallops, etc.

2. Place one clam in a dissection tray with the hinge side facing away from you. Also have the hinge closer to the left end (see diagram). Insert the scalpel between the upper shell and its underlying mantle (the layer of skin loosely attached to the shell). There will be two muscles holding the shells together: the anterior adductor and posterior adductor muscles. Sever both muscles as close as possible to where they connect to the shell. Once they are severed, open the upper shell wide or break it off where it hinges to the lower shell. At the right end of the clam (toward the rear), the mantle forms two apertures. The lower one is the **incurrent siphon,** which allows water to flow in. The upper one is the **excurrent siphon** which allows water to flow out. Between the main body of the clam and the sheet-like mantle will be two leaf-like **gills** lying side by side (there are two gills on the other side). Cilia on the gills generate a current which pulls water into their hollow interior. Food particles collect on mucus covering the surface of the gills. Cilia sweep the mucus containing trapped food (conveyor belt style) to the edge of the gills where it rides another cilia conveyor belt along the edge of the gill forward (toward the left) to the **palps.** The palps look like miniature versions of the gills. The palps (also ciliated) transfer the food to the adjacent mouth. Oxygen in the water passing through the gills diffuses into the blood stream that is circulates through the gills. The water that enters the gills moves upwards to the base of the gills and is directed rearward toward the excurrent siphon where it is expelled. The single **foot** is below the mouth and palps and can protrude out of the two shells down and forward. The muscular foot can extend forward into the substrate (mud or sand). Blood is pumped into the foot causing it to swell against the substrate, getting a firm grip. The foot then shortens by contracting longitudinal muscles which pulls the clam forward.
3. Observe a variety of bivalve shells.

C. Class Gastropoda

1. Gastropods are mollusks with one shell (usually spiraled). One can generally call them "snails," but each group usually has its own special name. Examples include the snails, whelks, cones, conches, periwinkles, cowries, etc. There are a few exceptions with no shells: These naked gastropods are the nudibranchs and land slugs.
2. Examine a large leopard slug and closely observe several external features. On the head there are two pair of tentacles; a longer **upper tentacles** tipped with eyes and a shorter **lower tentacles** for sensing its chemical environment. Behind the head there is a thickened saddle of skin called the **mantle** (this structure is what secretes the shell in most gastropods). The belly of the slug, called the **foot**, secretes a carpet of slime on which it crawls, using complex waves of muscular contractions. On the right side of the mantle is a hole that can change shape. This is called the **pneumostome**. The lining of the cavity serves as a lung for gas exchange.
3. Observe a variety of gastropod shells.

D. Class Cephalopoda

1. Cephalopods ("head-foot") are the most intelligent and structurally complex mollusks. Examples include chambered nautilus, cuttlefish, squid, and octopus.
2. Observe the nautilus shell and the preserved squid or octopus.

LABORATORY 19

PHYLUM ARTHROPODA

MATERIALS

Suggested vendor: Carolina Biological Supply Company

- Stereoscope
- Basic dissection kit / latex gloves
- Preserved insects from three major insect orders: Orthoptera, Coleoptera, and Lepidoptera (grasshopper, ground beetle, and butterfly)
- Preserved large millipede
- Preserved large centipede
- Preserved large spider
- Preserved crayfish

PREPARATION

Make sure you have read Chapter 25 "Phylum Arthropoda" in *The Riot and the Dance* (pp. 291–314). Arthropods are not only extremely diverse, they are also by far the largest phylum. They constitute over 80% of all animal species. In this lab, I will emphasize Class Insecta but we will still look at the main representatives of familiar groups of arthropods. The huge disparity in morphology and natural history is staggering when considering the entire phylum. Enjoy!

OBJECTIVES

i. Know the general anatomy of insects, millipedes, centipedes, arachnids (spider), and decapods (crayfish). (Figures 25.7, 25.10, and 25.38 in the student book)

ii. Be able to identify by sight three major orders of insects, their distinguishing characteristics, and their basic natural history (Keep in mind there are about 30 orders and many are very big. I'm trimming it down to three due to time limitations).

iii. Be able to identify by sight the major groups of non-insect arthropods: Class Diplopoda (millipedes), Class Chilopoda (centipedes), Class Arachnida (eight-legged arthropods represented by a spider), and the Subphylum Crustacea (represented by a crayfish).

EXERCISES

A. Definitions

Use the textbook glossary to define the following terms.

tegmina: ______________________________

elytra: ______________________________

proboscis: ______________________________

chelicerae: ______________________________

pedipalps: ______________________________

spinnerets: ______________________________

B. Class Insecta

1. Insects are the biggest group of arthropods having three body regions: head, thorax, and abdomen. Two pairs of wings arise from the thorax. A pair from the mesothorax and metathorax: hence four wings. Three pairs of legs arise from the thorax. A pair of legs arise from three thoracic segments: **prothorax**, **mesothorax**, and **metathorax**: hence, six legs. The abdomen has a variable number of segments depending on the kind of insect. Basic insect anatomy will be covered as we survey three insect orders.

Order Orthoptera *(grasshoppers, katydids, and crickets)*

2. Place a preserved grasshopper in a dissection tray and refer to Figure 25.10 (p. 298 in the textbook). Lift up the first pair of wings, which are leathery, narrow, and not used for flying. They are called **tegmina** and cover and protect the membranous hind wings. Tegmina have straight, parallel veins. That's why this order is called Orthoptera, meaning "straight wing" (*ortho* means "straight" and *ptera* means "wing"). Also examine the obvious and large jumping hind legs. This is the most telltale feature of this order. Pull open the hind wings which are used for flying. Note how they are pleated and fold up like a Japanese fan.

   What thoracic segment are the jumping hind legs on? ________________

   ____

Order Coleoptera *(beetles)*

3. Place a preserved ground beetle in a dissection tray. Pry up the first pair of leathery or rigid wing covers. These are the forewings called **elytra** (not used in flying) and form a protective sheath for the membranous hind wings. This is why they are called Coleoptera "sheath wings." When the two elytra are together at rest they form a straight seam down the middle of the back. There are over 350,000 species in this order with amazing variety in size, shape, color, and natural history.

4. What thoracic segment are the elytra attached to (hint: the two pairs of wings arise from the mesothorax and metathorax)? ________________

____

5. Observe the beetle from above and note the segment between the head and the base of the elytra. Many people learning insect anatomy assume that it's the thorax—but it's only the first segment of the thorax. What is that segment called? ________________________________

6. The elytra cover the entire abdomen and which two segments of the thorax?

________________________ and ________________________

7. Observe the underside of the beetle. If legs are arising from it, it is thorax. Note how big the thorax is and how short the abdomen is.

Order Lepidoptera ***(butterflies and moths)***

8. Examine a preserved butterfly with spread wings. Both pairs of wings are usually beautifully patterned. Examine the wings under the stereoscope and zoom in. Look at the differently colored tiny scales. The scales are like pixels and collectively they form the pattern of the wings. This is why they are called Lepidoptera meaning "scale wings." Examine the head from the side under the stereoscope. The mouthparts are coiled up like a party horn. This is called the **proboscis** and is a straw for sucking up nectar or other liquid foods. When feeding, the butterfly uncoils it using tiny muscles.

C. **Class Diplopoda** ***(the millipedes)*** **and Class Chilopoda** ***(the centipedes)***

These two classes of arthropods have two body regions: the **head** and the **trunk**.

1. Place a preserved millipede and centipede in a dissection tray. Examine the underside of a millipede using a stereoscope. How many pairs of legs does it have per body segment (excepting the first few segments)? ____________

   Based on your reading, can millipedes sting? ____________________

   Are they herbivores or predators? ____________________________

2. Examine the undersides of a centipede using a stereoscope. How many pairs of legs does it have per body segment? ______________________

   Based on your reading, can centipedes sting? ______________________

   Look at the underside of the head area. What do you see curving forward from the first trunk segment? ______________________

   Are they herbivores or predators? ______________________

D. Class Arachnida *(eight-legged arthropods)*

This class of arthropods has two body regions: the cephalothorax (head-thorax) and abdomen. If you're creeped out over examining the spider, take courage...it's dead. Besides, you can use a dissection needle to poke it and point things out.

1. Place a preserved spider in a dissection tray. Refer to Figure 25.38 (p. 314 in the text). Examine the body region with all the legs. What is this body region called? ______________________

2. Examine the face of the spider head on under the stereoscope. Note a pair of short vertically arranged appendages in the center of its face. At the tips of these are the **fangs.** These fang-tipped appendages are called the ______________________. Just to the side of these is a pair of ______________________ that look like a miniature pair of legs.

   How many pairs of walking legs are there behind these? ______________

3. Now closely observe the opposite end; tip of the abdomen. Note a set of conical **spinnerets** (two to eight; usually six) at the tip of its abdomen. These are responsible for ejecting silk with which they weave a number of useful products.

E. Subphylum Crustacea *(crustaceans)*

This group of arthropods is highly variable. We will examine a representative of a well-known order called Decapoda, which includes lobsters, crayfish, crabs, and shrimp.

1. Place a preserved crayfish in a dissection tray. Refer to Figure 25.7 (p. 296 in the textbook). Examine the body region with all the legs. What is this body region called? ______________________________

   Examine the face of the crayfish under the stereoscope. Under high power examine a compound eye. Why is it called a compound eye? ____________

   Note a pair of short pair of antennae that are forked.
   These are called ______________________________.

   Examine the other pair of antennae. How to they differ from the first pair?
   Count all the legs including the big pincers. How many? ______________

   What are the big pincers called? ______________________________

2. Using scissors, cut away the left side of the **carapace** (the large saddle-shaped exoskeleton plate covering the cephalothorax). You will see several whitish feathery gills. You will discover that the gills are attached to the bases of the walking legs if you carefully pull one leg off at its base.

3. Examine the abdomen's underside. You'll note pairs of small appendages called ______________________________.

   How many pairs are there? __________

   On the end of the abdomen there is a tail used for propulsion composed of several flat plates.
   The single, central plate is called the ___________________________ and the paired plates on the sides are called ___________________________.

LABORATORY 20

PHYLUM ECHINODERMATA

MATERIALS

Suggested vendor: Carolina Biological Supply Company

- Stereoscope
- Glass dish
- Dissection tray / latex gloves
- Preserved sea star
- Preserved brittle star
- Preserved sea urchin
- Sea urchin test (shell)
- Preserved sea cucumber

PREPARATION

In preparation for Lab 20 read carefully Chapter 26 "Phylum Echinodermata" in *The Riot and the Dance* (pp. 319–325). Echinoderms are simply mind-bogglingly bizarre compared to most animals, even invertebrates. They seem to break all conventions of typical animal anatomy. In this lab I will emphasize Class Asteroidea (sea stars) but we will briefly look at the main representatives of the other echinoderm classes. Enjoy!

OBJECTIVES

i. Know the general characteristics of echinoderms.

ii. Know the general anatomy of a sea star, both internal and external. (Figure 26.2 in the textbook)

iii. Be able to identify by sight four major classes of echinoderms, their distinguishing characteristics, and their basic natural history.

EXERCISES

A. Definitions

Use the textbook glossary to define the following terms.

central disc: ____________________

arms: ____________________

ambulacral grooves: ____________________

water vascular system: ____________________

pyloric stomach: ____________________

cardiac stomach: ____________________

papulae: ____________________

pedicellariae: ____________________

B. Class Asteroidea *(the sea stars)*

1. Place a preserved sea star in a dissection tray. The underside is called the **oral surface** and the top side is the **aboral surface.** I will use these terms in directing you where to look. Note the **arms** radiating out from the **central disc.** How many arms does your specimen have? ____________

2. Look at the oral surface. What are the grooves running the length of each arm called? ____________________

3. Examine the grooves under the stereoscope. They are packed full of tiny macaroni-like projections called ________________________________.
   Based on Figure 26.2, these are part of the hydraulic system called the
   ________________________________ system.
   What part of this system contracts to extend these tube feet using water pressure? ________________________________

4. Look at the aboral surface of the central disc. Note a whitish, rock-hard round patch a little off-center. That is the **sieve plate** which is the opening into the water vascular system. Microscopic pores on this plate filter the sea water before it enters the **stone canal**. Find the tube feet, ampullae, sieve plate, and stone canal, but don't bother trying to find the rest of the water vascular system (unless you have oceans of time).

5. Cut the tip off (about a ¼ of an inch from the end) of one arm. Using scissors (starting where the tip was cut off), cut the aboral surface exposing the inside of the arm. After you've seen the sieve plate, cut the aboral surface off the central disc. Try to cut around the sieve plate so that it isn't removed with the skin.

6. Place a patch of skin in a small dish of clean water about a half inch deep. Examine the patch of skin under the stereoscope. Note the rounded white spines that protrude from the skin. These are **spines.** That's why these creatures are called echinoderms: *echino* means "spiny" and *derm* means "skin." Much smaller than the spines are soft finger-like projections scattered over the surface of the skin, these are **papulae** or skin gills. You may see tiny strange pincer-like structures on short stalks. These are called **pedicellariae.**

7. Inside the arm you'll see a pair of muddy greenish glands taking up a good deal of the room inside the arm. These are **digestive glands.** The digestive glands dump their digestive enzymes into the upper **pyloric stomach** in the central disc. Remove the digestive glands and you will see two long rows

of many tiny, rounded sacs called **ampullae** on the inside of the bottom of the arm. These are directly opposite the tube feet that stick out from the ambulacral grooves on the other side.

8. Examine the middle of the oral surface. This opening is the **mouth** which immediately leads to the **cardiac stomach** (the pyloric stomach is directly above it). In many sea stars, the cardiac stomach is extruded out through the mouth and into the prey where it is partially digested outside the body using all the enzymes from the digestive glands. A short tube connects the pyloric stomach to the **anus** (which is on the aboral surface but very difficult to find).

C. Class Ophiuroidea *(the brittle stars)*

Place a preserved brittle star in a glass dish to observe under the stereoscope. Describe in a few sentences how its central disc and arms differ from the sea star's.

____

____

____

____

D. Class Echinoidea *(the sea urchins, sea dollars, etc.)*

1. Place a preserved sea urchin in a glass dish to observe under the stereoscope. Does this echinoderm have a central disc and arms? ____________________
   What are its spines like when compared to the sea star spines?

   ____

   ____

2. Hold a sea urchin test up to the light and observe it closely. You will see light shining through pin holes. The pin holes form rows (like longitudinal

lines on a globe) from the center of the aboral surface to the center of the oral surface. These pin holes are where the tube feet emerge from the test. These are called ambulacral areas and correspond to the ambulacral grooves of sea stars.

What are the feeding habits of most sea urchins?____________________

____

What apparatus does it use to cut and chew its food? ________________

____

E. Class Holothuroidea *(the sea cucumbers)*

Place a preserved sea cucumber in a glass dish or tray. Does this echinoderm have a central disc and arms? ____________

Does it have hard spines projecting from its body? ______________________

What are the bushy things surrounding its mouth? ______________________

What are they used for? ____

You'll notice longitudinal rows (usually five) running from the mouth (in front) to the anus (in back). What are those rows composed of? __________________

LABORATORY 21

PHYLUM CHORDATA 1

THE FISHES

MATERIALS

Suggested vendor: Carolina Biological Supply Company

- Stereoscope
- Dissection tray / latex gloves
- Preserved lamprey (Agnatha)
- Preserved dogfish shark (Class Chondrichthyes)
- Preserved perch (Class Osteichthyes)

PREPARATION

Make sure you have read up to Class Amphibia in Chapter 27 "Phylum Chordata" in *The Riot and the Dance* (pp. 327–337). Chordates are the animals we are most familiar with (for the most part). The vast majority of them are vertebrates, which includes the fish, amphibians, reptiles, birds, and mammals. In the first Chordate lab we only have time to briefly survey three groups of fish. In the second Chordate lab we will briefly survey three classes of tetrapods (four-legged vertebrates): Class Amphibia, Class Reptilia, Class Aves, and Class Mammalia. Even though it's a flyby we can at least see some key features of each and make important distinctions between them.

OBJECTIVES

i. Know the general characteristics of chordates.

ii. Know the general characteristics of the Agnatha, Class Chondrichthyes, and Class Osteichthyes. (To simplify things, I have lumped two classes of jawless fishes together into the informal group Agnatha.)

iii. Be able to identify by sight these three fish groups, their distinguishing external characteristics, and representative examples of each (for example, Class Chondrichthyes: sharks, sawfish, skates, and rays).

EXERCISES

A. Definitions

Use the textbook glossary to define the following terms.

dorsal nerve cord: ____________________

notochord: ____________________

pharyngeal slits: ____________________

dorsal fin: ____________________

caudal fin: ____________________

pectoral fins: ____________________

pelvic fins: ____________________

anal fin: ____________________

gill slits: ____________________

gill rakers: ____________________

operculum: ____________________

B. Agnatha *(the jawless fishes)*

1. Place a preserved lamprey in a dissection tray.

   Does it have a hinged jaw to open and close the mouth?________________

Note the keratinized teeth inside the suction-cup-like mouth which they use to grasp onto the skin of a fish.

2. Count the pairs of gill slits. How many are there? ____________
3. Examine the fins. Are they paired? ___________
4. Are all lampreys ectoparasitic on other fish? ___________
5. What is the other jawless fish mentioned in the textbook? ______________

C. Class Chondrichthyes *(the cartilaginous fishes)*

1. What does *khondros* mean in Greek? ______________________________

   What does *ikthus* mean in Greek? __________________________________

   What structure do you think this name refers to in these fish?

   ____

2. Place a preserved dogfish shark in a dissection tray. Let me ask you a stupid question. Does it have a hinged jaw to open and close the mouth?__________

   Open its jaw and examine its many sharp teeth arranged in rows. It may have a cartilage skeleton, but what do you think its jaw is made of? _____________

3. Feel the texture of the skin (without gloves). *You can wash your hands afterwards if you like.* Cut off a small piece of skin from its side and examine the scales under a stereoscope. Zoom in to the highest power.

   What do the scales look like? ______________________________________

   What is the name of this type of scale? _____________________________

4. How many pairs of gill slits does it have? ____________

5. Draw a sketch of this shark (side view) and label its gill slits and the following fins: dorsal, caudal, pectoral, pelvic, and anal.

6. List a few different fishes that belong to Class Chondrichthyes besides sharks.

____

____

____

____

D. Class Osteichthyes *(the bony fishes)*

1. What does '*ostoun*' mean in Greek? ______________________

   What structure do you think this name refers to in these fish?

   ____

2. Place a preserved perch in the dissection tray. Note the difference in the placement of the mouth. Where is it positioned compared to the shark mouth?

   ____

3. Cut a small piece of skin and examine the scales under the stereoscope. Briefly describe the difference in size and shape compare to the shark skin.

____

____

____

4. Do you see gill slits? __________

   What is the structure called covering the gills? ______________________

   Do sharks have this structure? ____________

5. Cut off this covering with scissors to reveal the gills. Cut off one of the gill arches (Figure 27.13 in the textbook) by cutting the bases at both ends. Place it in a glass dish submerged in water and examine it under the microscope. Note the longer thin frills on the convex surface of the gill arch. What are these called and what are they used for? __________________

   ____

   Note the shorter, stiffer, comb-like projections on the concave surface of the gill arch. These are **gill rakers**. What are they used for? ____________

   ____

6. Draw a sketch of this bony fish (side view) and label its operculum and the following fins: the dorsal, caudal, pectoral, pelvic, and anal fins.

LABORATORY 22

PHYLUM CHORDATA 2

AMPHIBIANS & REPTILES

MATERIALS

Suggested vendor: Carolina Biological Supply Company.

If live amphibian or reptile pets can be brought in, they are always better than preserved examples (unless they're moving too much).

- Compound microscope
- Frog skin microscope slide
- Stereoscope
- Dissection tray / latex gloves
- Preserved salamander—*Necturus* (Class Amphibia)
- Preserved frog—*Rana* (Class Amphibia)
- Preserved lizard—*Anolis* (Class Reptilia)
- Preserved turtle (Class Reptilia)

PREPARATION

Make sure you have read the two tetrapod classes Amphibia and Reptilia in Chapter 27 "Phylum Chordata" in *The Riot and the Dance* (pp. 337–352). People are generally familiar with Class Aves (birds) and Class Mammalia (mammals), so I will omit them in the lab and let you read about them in the textbook. These brief overviews are not meant to be the 'beginning and end' of all that's worth learning about these vast groups. In formal education there are always time limits. Therefore, I encourage you to let these brief introductions be not only a foundation but also a springboard into a lifelong exploration of one or more of these animal groups—not just vertebrates, but anything that we've covered up to now. Many non-biologists enjoy nature and turn a fascination of a certain group into a hobby like "birding" or "herping" or whatever. Don't let formal learning kill curiosity.

OBJECTIVES

i. Know the general characteristics of chordates.
ii. Know the general characteristics of Class Amphibia and Class Reptilia.
iii. Be able to identify by sight these two tetrapod classes, their distinguishing external characteristics, and representative examples of each (for example, Class Amphibia: frogs, toads, salamanders, and caecilians).

EXERCISES

A. Definitions

Use the textbook glossary to define the following terms.

herpetology: ______________________________

ectothermic: ______________________________

spermatophore: ______________________________

amplexus: ______________________________

plastron: ______________________________

carapace: ______________________________

osteoderms: ______________________________

B. Class Amphibia *(the amphibians)*

1. Place a preserved salamander (*Necturus*) and frog in a dissection tray. Closely observe the skin and toes on both closely. Is the skin smooth or scaly?

   ____

   Do their toes have claws? __________

   Note the feathery gills on the neck region of the salamander. These are usually a larval salamander characteristic but some species retain their gills to adulthood. Besides the gills, what does the salamander have that the frog doesn't?____

   What does the frog have that the salamander doesn't?__________________

   ____

   Based on your textbook reading, what is the difference between how male frogs fertilize eggs and how male salamanders fertilize eggs?

   ____

   ____

   ____

   ____

2. Place the frog skin slide under the microscope. On scan, then low, then high power note the large round openings. Some will be connected to the surface of the epidermis with a duct (tube). These are **mucus glands.** They secrete mucus into the lumen (round opening). The mucus is forced up through the duct and out onto the surface of the skin to keep it moist. Mucus covered skin allows it to absorb oxygen which diffuses through the thin epidermis and into the blood capillaries of the dermis. The round part of each gland is in a lighter layer called the **dermis.** The upper darker pink to purplish layer surrounding the ducts is the **epidermis.** Compared to

our epidermis amphibian skin is much thinner. The area below the dermis filled with pink parallel wavy lines is a **muscle layer.**

C. **Class Reptilia *(the reptiles)***

1. Place a preserved lizard (*Anolis*) in a dissection tray but also keep the salamander on the tray for comparison. Many people confuse lizards with salamanders. I hope to remedy this problem with a few simple observations. Closely observe (with the stereoscope if you wish) the skin and toe tips on both. Is the lizard skin smooth or scaly?______________

   Do lizard toes have claws? _________

2. Describe any other obvious external differences you see between the lizard and salamander.

Lizard	Salamander
__________________________	__________________________
__________________________	__________________________
__________________________	__________________________
__________________________	__________________________
__________________________	__________________________
__________________________	__________________________
__________________________	__________________________
__________________________	__________________________

   Based on your textbook reading, what reproductive structure do male lizards have that male salamanders don't? ______________________

3. Place a preserved turtle in a dissection tray. Observe the top shell (**carapace**) and the bottom shell (**plastron**). Note the large tile-like plates on both these shells. These large scales are called **scutes.** Under these scutes are fused osteoderms. Osteoderms are formed from dermis (part of the skin) that has turned into bone.

Do the legs have scales? ________

What parts of its skin doesn't have scales? ______________________

____

Do the toes have claws? ___________

4. Place a preserved non-venomous snake in a dissection tray. Note the differences in size and shape of the scales on its body. What are snake belly scales shaped like? ______________________________

5. Based on your textbook reading, what reproductive structure do both male lizards and snakes have in common? ______________________

   What do male turtles have instead? ______________________

6. Examine the lizard eye and ear areas compared to the same areas in the snake. If you found a legless lizard (they do exist) how could you tell that it wasn't a snake? In other words, what two things would they have that snakes don't?

   a. __________________________

   b. __________________________

LABORATORY 23

KINGDOM PLANTAE 1

THE MOSSES & FERNS

MATERIALS

Suggested vendor: Carolina Biological Supply Company.

- Compound microscope
- Stereoscope
- Fresh moss with sporangia
- Moss sporangium (capsule slide)
- Moss protonema slide
- Moss antheridia slide
- Moss archegonia slide
- Fresh fern leaves with sori
- Fern leaflet with sporangia slide
- Fern prothallus slide

PREPARATION

Make sure you have read through "The Ferns" of Chapter 28 "Kingdom Plantae" in *The Riot and the Dance* (pp. 367–374). Because plants don't move around like animals they are sometimes overlooked as pretty props on the stage of life. I hope to change your perspective. Yes, they definitely add verdant beauty to our gardens and homes, but once their subtle, intricate way of life is revealed it makes them become objects of our focused interest as well as a source of beauty. There are four familiar phyla in the Plant kingdom: Phylum Bryophyta (mosses), Phylum Pterophyta (ferns), Phylum Coniferophyta (conifers), and Phylum Anthophyta (flowering plants).

OBJECTIVES

i. Know the general characteristics of plants.
ii. Know the general alternation of generations life cycle of plants (Figure 28.1 in the textbook).
iii. Understand the moss life cycle (along with its reproductive structures) and how it fits the alternation of generations cycle (Figure 28.2).
iv. Understand the fern life cycle (along with its reproductive structures) and how it fits the alternation of generations cycle (Figure 28.9).

EXERCISES

A. Definitions

Use the textbook glossary to define the following terms.

sporophyte: ____________________

sporangium: ____________________

spores: ____________________

protonema: ____________________

gametophyte: ____________________

antheridium: ____________________

archegonium: ____________________

frond: ____________________

rhizome: ____________________

pinnae: ____________________

sorus (sori):____________________

annulus: ____________________

prothallus: ____________________

B. Phylum Bryophyta *(the mosses)*

1. Place a fresh piece of moss in a glass dish. Note the lush green leafy carpet with a number of thin leafless stalks sticking up and oval knobs on the tips (Figure 28.4 in the textbook). The leafy green carpet is the **gametophyte** generation. What is the ploidy of the gametophytes? ____________

   The leafless stalks with knobs are the **sporophyte** generation. What is the ploidy of the sporophytes? ____________________

   What is the knob on top called? ____________________

   What type of cell division occurs within the knob to make spores?

   ____

2. Look at the moss sporangium slide with the naked eye first, then place it under the microscope and view it on scan then low power. See the tiny haploid **spores** within it. When they are haploid and ready the cap of the sporangium pops off and the spores are released. Some chance to land on a suitable moist surface where they can grow.

3. View the moss protonema slide under the microscope. This is what the spores grow into. They look like filamentous green algae. What is their ploidy?

   ____

   These eventually grow into gametophytes (some male, some female).

4. View the male gametophyte (look at it with the naked eye first). Position it on scan power so that you see the top of the tiny leafy gametophyte. You'll see

oval shaped structures filled with hundreds of moss sperm (move to higher power when you find these). These sperm containers are called **antheridia.**

5. View the female gametophyte (look at it with the naked eye first). Position it on scan power and position it so that you see the top of the tiny leafy gametophyte. You'll see narrow vase-shaped structures each containing one egg (move to higher power when you find these). These egg containers are called **archegonia** (Figure 28.3 in the textbook). Water (precipitation) breaks the antheridia open releasing the sperm. Some sperm swim through the water film to archegonia. If a sperm swims down the neck and fertilizes the egg it becomes a diploid **zygote.** The zygote is the beginning of the new sporophyte generation. It grows straight up out of the archegonium and forms the stalk and knob (the mature sporophyte with its sporangium)

C. **Phylum Pterophyta** ***(the ferns)***

1. Place a fresh piece of fern leaflet (bottom-side up) under the stereoscope. Examine the spots called sori on the underside of the leaflet. The entire fern plant (roots, rhizomes, and leaves) is the sporophyte generation.
2. Look at the fern sporangium slide with the naked eye first (it will be looking at the sorus from a side view (Figure 28.8 in the textbook). Under the microscope you will see that the sorus is a cluster of sporangia. Examine one sporangium at high power. See the haploid spores within it and a ring of cells called an annulus surrounding the sporangium. What type of cell division occurs within the knob to make spores? ____________

   When the spores are haploid and ready for release, the annulus curls open and catapults the spores violently into the air. Some chance to land on a suitable moist surface where they can grow into a fern **gametophyte** (about the size of a flake of dandruff). What is the ploidy of the gametophyte?

   ____________________.

The fern gametophyte is called a **prothallus.** There aren't separate male and female prothalli. Rather both antheridia and archegonia are on the same prothallus.

3. Place a fern prothallus slide under the microscope. You'll see little dark bumps in the middle of this heart-shaped prothallus (near where the two lobes of the heart meet). Some will be **antheridia** and others will be **archegonia.**

   In a similar fashion as in the moss life cycle, the sperm is released from the antheridia and fertilize the egg in the archegonia to produce a diploid **zygote.** The zygote grows out of the prothallus and grows into a tiny sporophyte (Figure 28.11 in the textbook). Eventually it grows into the mature fern plant (the adult sporophyte).

4. Examine a fern plant and identify the following parts: the **rhizome, roots, fronds** (leaves) and its **leaflets** (pinnae), and **petioles** (Figure 28.6 in the textbook).

LABORATORY 24

KINGDOM PLANTAE 2

THE CONIFERS & FLOWERING PLANTS

MATERIALS

Suggested vendor: Carolina Biological Supply Company.

- Compound microscope
- Stereoscope
- Male pine cone slide
- Female pine ovule (with megagametophyte) slide
- Lily anther cross section (mature pollen)
- Lily ovary (with mature ovules) cross section slide
- Fresh pine twig with female cones on it
- Fresh tulip or lily flower

PREPARATION

Make sure you have read Phylum Coniferophyta through Phylum Anthophyta of Chapter 28 "Kingdom Plantae" in *The Riot and the Dance* (pp. 374–385). These two phyla are the most familiar to us since these are the vast majority of the plants we see in the woods, our gardens, parks, and in our homes. This familiarity sometimes gives us the idea that we know them better than we do. I think this lab will reveal to us how little we actually know about their life cycle.

OBJECTIVES

i. Know the general characteristics of plants.
ii. Know the general alternation of generations life cycle of plants (Figure 28.1).
iii. Understand the conifer life cycle (along with its reproductive structures) and how it fits the alternation of generations cycle.
iv. Understand the flowering plant life cycle (along with its reproductive structures) and how it fits the alternation of generations cycle.

EXERCISES

A. Definitions

Use the textbook glossary to define the following terms.

male cone: ____________________

microsporangium: ____________________

microgametophyte (pollen): ____________________

female cone: ____________________

ovuliferous scales: ____________________

ovule: ____________________

megagametophyte: ____________________

sepals: ____________________

petals: ____________________

stamens: ____________________

anthers: ____________________

filaments: ____

pistil: ____

stigma: ____

style: ____

ovary: ____

B. Phylum Coniferophyta *(the conifers)*

1. To make a connection to the life cycle, the entire pine tree is the **sporophyte**. Observe a fresh twig with male cones on it (Figure 28.16 left in the textbook) which is part of the sporophyte.
2. With a scalpel slice one of the cones lengthwise and place it under the stereoscope. Zoom in on several small scales bearing yellowish sacs of pollen (if they are mature). These sacs are called **microsporangia.** Inside these sacs diploid cells go through meiosis to produce four **microspores** (it follows the general life cycle perfectly). The only twist is the prefix "micro" which denotes "male". Each microspore goes through a couple rounds of mitosis to make a very tiny **microgametophyte** called a pollen grain. What is the ploidy of the microgametophytes? ____________________
3. Observe a fresh twig with female cones on it (Figure 28.16 right in the textbook) which is part of the sporophyte. The big woody scales are called ovuliferous scale (ovule-bearing scale) Each scale has two ovules on it.
4. Place the slide of the female pine ovule (with megagametophyte on the compound microscope. On scan power observe the ovule lying on top of the ovuliferous scale. This ovule (immature seed) has been sectioned so as to reveal the **megagametophyte** (female gametophyte). On the side facing the cone's axis, there is a micropyle or opening into the ovule. The micropyle is where the pine pollen must drift by the wind for pollination to be achieved.

The pollen grain produces a pollen tube that grows to the megagametophyte within the ovule. The megagametophyte has 2 or 3 archegonia each containing an egg. Sperm from the pollen grain travels through the pollen tube and fertilizes the egg which becomes the diploid zygote. The zygote grows into a tiny sporophyte embryo and stops development within the ovule. It is now a seed. It eventually flutters down to the ground from the cone and awaits the right conditions to germinate into a pine seedling. It follows the life cycle of alternation of generations perfectly.

C. Phylum Anthophyta *(the flowering plants)*

1. Cut a fresh lily flower with a scalpel, making sure to cut the entire pistil lengthwise as well. Refer to Figure 28.28 in the textbook to identify the following floral parts: sepals, petals, stamen (filament & anther), pistil (stigma, style, and ovary).
2. Sketch the flower you cut and label these parts. Also include in your drawing the many ovules within the ovary, since you sliced it open lengthwise.

3. Place a slide of a mature anther under a compound microscope. Examine one of the four pollen sacs in the anther cross section (scan → low → high). Observe the pollen grains (**microgametophytes**) within the sacs. Earlier diploid cells underwent meiosis to form **microspores.** Each microspore produced a **microgametophyte** through a couple rounds of mitosis. Inside the pollen grain is a cell that divides to produce two sperm. When wind or animals (mostly insects) whisk pollen away from the anther and deposit it on a stigma of the same species (usually not the same flower) it is called ____________________ .

4. Now look at a slide of a cross section of a mature lily ovary. Inside the ovary are small **ovules** connected to the center area of the ovary. The ovary is partitioned into three sections. Within each ovule is a **megagametophyte.** There are so few cells that there is no archegonium, but there is an egg near the opening of the ovule called the micropyle. Don't worry if you can't see the egg or the megagametophyte. Just know what the ovules look like under the microscope. Hopefully you can see the micropyle after you switch to high power.

5. After pollination, the pollen grain germinates on the stigma. The pollen tube grows down through the stigma, style, and ovary wall to an ovule. It grows up into a micropyle and stops near the egg. One of two sperm travels down the pollen tube and eventually arrives at the egg. The sperm _ the egg producing a diploid zygote which then grows into a tiny sporophyte embryo. After growing a little, it stops development within the ovule. The ovule is now a **seed** and must receive the proper conditions to germinate into a seedling.

LABORATORY 25

ECOLOGY

MATERIALS

- A computer to watch an online video

PREPARATION

Make sure you have read Chapter 29 "The Basics of Ecology" in *The Riot and the Dance* (pp. 389–402). Ecology is simply the study of living creatures and their interactions with their environment (which includes other living creatures). There are many aspects to ecology which can't even be summarized in one short lab. However, we will watch a segment of a nature documentary to see how many ecological concepts can be introduced.

OBJECTIVE

i. Understand and be able to identify certain ecological concepts by giving examples from the documentary.

EXERCISES

A. Definitions

Use the textbook glossary to define the following terms.

ecology: ____________________

abiotic factors: ____________________

biotic factors: ____________________

population: ____________________

community: ____________________

ecosystem: ____________________

habitat: ____________________

microhabitat: ____________________

ecological niche: ____________________

predator-prey relationship (not in glossary): ____________________

mutualism: ____________________

commensalism: ____________________

parasitism: ____________________

competition: ____________________

interspecific competition: ____________________

intraspecific competition: ____________________

exponential growth: ____________________

logistic growth: ____________________

carrying capacity: ____________________

food chain: ____________________

food web: ____________________

trophic pyramid: ____________________

trophic level: ____________________

decomposers: ____________________

detritovores: ____________________

B. Video

Visit http://logospressonline.com/content/RiotLinks.pdf, and click on the link for the Laboratory 25 video. This video is one episode of a BBC series, *Life in Cold Blood*, called "Dragons of the Dry." The entire series is about amphibians and reptiles. Although it has a strong evolutionary bias, the footage is spectacular and can visually showcase quite a number of ecological concepts. Fill out the worksheet below as you watch this documentary.

C. Dragons of the Dry

1. Lace monitors lay their eggs in the nests of termites. By laying the eggs there, what two abiotic factors are they giving to their young to encourage hatchling success?

   a. ______________________________

   b. ______________________________

2. What biotic factor is the nest protecting the eggs and hatchlings from?

   ____

3. Anoles head bob, do push-ups (press-ups), and display their dewlaps to defend their territory. What kind of competition is this an example of?

   a. Interspecific competition

   b. Intraspecific competition

4. Give two additional behaviors of the Jackie Dragon that are also territorial displays that show this kind of competition.

   a. ______________________________

   b. ______________________________

5. Most animals are in predator-prey relationships. Even tiny lizards can be predators. Give one example: ____

   What does it eat?____

6. The Tokay Gecko is a predator of ____________________________.

7. In a predator-prey relationship, the prey species often have camouflage that keeps them from being detected. What three characteristics make the Madagascan Leaf-tailed Gecko virtually invisible to a hungry predator?

   a. __________________________

   b. __________________________

   c. __________________________

8. How does the Pygmy Blue-tongued Skink keep from being detected by possible predators? ____

9. What is this sequence below called? ________________________________

Tree → plant hopper → Madagascan Day Gecko

10. The tree in this sequence is a:
    a. Producer
    b. Primary consumer
    c. Secondary consumer
    d. Tertiary consumer

11. The plant hopper is a:
    a. Producer
    b. Primary consumer
    c. Secondary consumer
    d. Tertiary consumer

12. Madagascan Day Gecko is not a predator (secondary consumer). What does it eat? ______________________________

13. What kind of relationship does it seem to have with the plant hopper?
    a. Mutualism
    b. Commensalism
    c. Parasitism

14. In a trophic pyramid, what trophic level is the tree in? ________________

15. In a trophic pyramid, what trophic level is the plant hopper in? __________